GENETIC ASSOCIATION STUDIES:

Background, Conduct, Analysis, Interpretation

GENETIC ASSOCIATION STUDIES:

Background, Conduct, Analysis, Interpretation

Mehmet Tevfik Dorak

Garland Science
Taylor & Francis Group
LONDON AND NEW YORK

Vice President: Denise Schanck
Senior Editor: Elizabeth Owen
Assistant Editor: David Borrowdale
Production Editor: Deepa Divakaran
Illustrator: Oxford Designers & Illustrators Ltd
Layout: Nova Techset Ltd
Cover Designer: AM Design
Copyeditor: Jo Clayton
Proofreader: Susan Wood
Indexer: Bill Johncocks

ISBN 978-0-8153-4463-6

Library of Congress Cataloging-in-Publication Data

Names: Dorak, M. Tevfik, author.
Title: Genetic association studies : background, conduct, analysis, interpretation / Mehmet Tevfik Dorak.
Description: New York, NY : Garland Science, Taylor & Francis Group, LLC, [2017] | Includes bibliographical references.
Identifiers: LCCN 2016030624 | ISBN 9780815344636 (alk. paper)
Subjects: | MESH: Genetic Association Studies | Genetic Phenomena | Molecular Epidemiology--methods
Classification: LCC QH437 | NLM QU 550 | DDC 576.5--dc23
LC record available at https://lccn.loc.gov/2016030624

Published by Garland Science, Taylor & Francis Group, LLC, an informa business, 711 Third Avenue, New York, NY, 10017, USA, and 2 Park Square, Milton Park, Abingdon, OX14 4RN, UK.

Printed in the United Kingdom
15 14 13 12 11 10 9 8 7 6 5 4 3 2 1

Visit our website www.garlandscience.com

Preface

This book has been written at probably the most challenging and exciting time in the recent history of genetics. The last decade has seen the completion of one revolutionary project after another, such as the ENCODE and NIH Roadmap Epigenomics projects. Both of these required the rewriting of a large proportion of the text, and the recent developments in translational bioinformatics forced the addition of a new chapter. *Genetic Association Studies* brings together in one place the most relevant aspects of genetics, environmental and genetic epidemiology, and statistics. Laboratory aspects of genetic association studies are also included. Although genome-wide association studies have changed the way we think about genetic associations, candidate gene studies and more basic genotyping methods are included to make sure that the book is relevant to students and researchers with limited resources. I hope every reader interested in genetic associations will find something useful here, and especially that students will benefit from having a single source covering different features of genetic associations. The book does not claim to be a comprehensive monograph on genetic epidemiology, but rather demystifies key concepts for those who have not received formal training in epidemiology or statistics.

Writing this book has felt at times like chasing a moving target, and sometimes I wondered how it was going to be possible to finish. It wouldn't have been possible without the understanding, encouragement, and tenacity of the staff at Garland Science, especially Liz Owen and David Borrowdale. A book incorporating information from different disciplines and written as a solo effort is bound to have imperfections. Expert reviewers worked very hard to help to improve the quality of information contained in the text. I am very much grateful to all reviewers. Any remaining errors are my responsibility. I must also thank my students, who have seen many different versions of this book and provided a lot of feedback, both knowingly and unknowingly. In particular, I am most grateful to my former graduate students, Amy E. Kennedy and Sandeep K. Singh, for their tireless efforts. My wife Christina suffered more than her fair share while I was trying to finish the book and I thank her for never-ending support and her encouragement in bringing this project to successful completion. Finally, I must express my gratitude to everyone who has contributed to my development, especially my family, teachers, mentors, and colleagues.

Mehmet Tevfik Dorak

Acknowledgments

The author and publisher would like to thank external advisers and reviewers for their suggestions and advice in preparing the text and figures.

Jennifer Barrett, University of Leeds, UK; Farren Briggs, Case Western Reserve University, USA; Corinne D. Engelman, University of Wisconsin, USA; Amy Kennedy, National Cancer Institute, USA; Christine Ladd-Acosta, Johns Hopkins Bloomberg School of Public Health, USA; Caroline Relton, Newcastle University, UK; Michael Routledge, University of Leeds, UK; James Tang, University of Alabama at Birmingham, USA; Dawn Teare, University of Sheffield, UK; Timothy Thornton, University of Washington, USA; Joe Wiemels, University of California, San Francisco, USA; Maurice Zeegers, University of Birmingham, UK.

Contents

Primer on Molecular Genetics

1

Genetic association studies are all about the association of various phenotypes with **genetic variants**. Genetic variation generates variability in phenotypes, including a spectrum of susceptibility to disease. The most common type of genetic variation is sequence differences between the copies of genes on different chromosomes. Genetic variation may also refer to changes in the copy number of short sequences, regions, or genes. Changes in the numbers or structures of chromosomes are gross types of variation, often incompatible with life, and are therefore rare. To be able to modify disease susceptibility, variants should affect either gene function or the structure of the encoded protein. A variant may modify disease risk via changes in genome biology (**Figure 1.1**). The most common trait or intermediate phenotype that is influenced by genetic variation is **gene expression**. This chapter provides background information on molecular genetics relevant to the understanding of the outcomes of genetic variation. Such information is important in variant selection for a genetic association study, for interpretation of results, and for designing **functional replication** experiments as follow-up studies. Very rare variations that are Mendelian disease-causing **mutations** are important in medical genetics. Other chapters providing background information for the statistical and epidemiologic aspects of genetic association studies follow this chapter.

1.1 Genetic Variation

Changes at the nucleotide level are most common

Genetic variation is due to mutation, with the most common mutation being a nucleotide substitution. When this type of variation becomes common in a population (that is, it is present in more than 1% of the population) it is called a **single nucleotide polymorphism (SNP)**. Variations in a single nucleotide are called **single nucleotide variations (SNVs)** and occur approximately once every 100 to 300 nucleotides in the human genome. The total number of SNVs is known to be above 160 million. Nucleotide variation may also be caused by insertion or deletion of nucleotides, called a **deletion or insertion polymorphism (DIP)** or an insertion-deletion (INDEL) polymorphism.

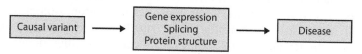

Figure 1.1 Intermediate mechanisms mediating a causal variant's effect on disease susceptibility. Genetic variants modify disease risk by causing changes in gene expression (most common), the splicing process, or protein structure.

Genes do not always exist as a single copy on the chromosome, but may be deleted or exist in multiple copies, known as a **copy number variation** (**CNV**). CNVs may have functional consequences on gene expression levels. Chromosomal trisomies result in multiple copies of each gene on that chromosome, but a CNV usually refers to a gene or a chromosomal segment rather than the whole chromosome. **Table 1.1** lists genetic variation types and their special features.

While nonpolymorphic positions have the same nucleotide in each chromosome in a population, alternative nucleotides at polymorphic positions are called **alleles**. The simplest variation is a SNP with two alleles, although SNPs may rarely have up to four alleles (there are four possible nucleotides in DNA). Since humans carry two chromosomes, a SNP with two alleles results in three possible **genotypes**, which in turn modify the phenotype (**Figure 1.2**). The phenotypes may be quantitative (measured characteristics like height, weight, blood pressure, or blood levels) or may be qualitative or discrete (such as the presence or absence of a trait, including a disease). Discrete traits relate to the presence or absence of disease, or having one blood group or another. A phenotype may show quantitative changes correlating with the number of copies of an allele in the genotype (a linear **gene-dosage effect**). Alternatively, the three genotypes may be collapsed into two groups by pooling two of them into one category depending on the **genetic risk model**. Such a model implies nonlinear effects of genotype upon phenotype.

Nucleotides form a double helix

The structure of DNA was described by Watson and Crick in 1953 as a double helix. It is composed of two intertwined chains of nucleotides—comprising nitrogenous bases attached to a sugar–phosphate backbone—that give rise to double-stranded DNA (dsDNA) (**Figure 1.3**). The component strands of the dsDNA are referred to as either sense or antisense strands, depending on which one is used to make RNA. It is the antisense strand that is used as a template for making RNA, resulting in RNA having the same sequence as the sense strand. The four nucleotide bases are adenine (A), cytosine (C), guanine (G), and thymine (T). The nucleotides are arranged, with the bases inside the helix, in a complementary fashion so that only A and T can pair with each other, and C with G, through hydrogen bonds. Whichever nucleotide is on the sense strand, its complementary nucleotide is on the antisense strand and pairs with it. Traditionally, the length of a DNA fragment is given in base pairs (bp) rather than nucleotides due to the double-helix structure of DNA. The helical structure of DNA, along with the proofreading mechanisms of DNA polymerase, enables DNA to replicate with high fidelity. Individual DNA strands are readily separated from each other—*in vivo* by enzymes and *in vitro* using heating or alkali treatments—after which the complementary nucleotides can join back together. DNA damage repair mechanisms are able to utilize the unzipping properties of the DNA molecule, using one strand as a template to repair the other, damaged strand.

Any of the four nucleotides A, C, G, and T may be the two alleles of a polymorphism, and the two alleles on corresponding positions in each pair of chromosomes make up genotypes. Both the metaphase chromosome that is usually used to illustrate a chromosome and the double-helix structure of DNA may create confusion when defining genotypes. Metaphase chromosomes are duplicated copies of long and short arms, which make them look like the letter X. They assume this shape only during metaphase in cell division. The double-stranded structure of the DNA molecule may also create confusion. As shown in **Figure 1.4**, when forming genotypes, nucleotides on the same strand (sense or antisense)

Table 1.1 Types of genetic variation

Genetic variation	Abbreviation	Features	Examples
Single nucleotide polymorphism	SNP	The variation involves one nucleotide. Most commonly a C-to-T substitution (C > T); also called a single nucleotide variant (SNV)	rs1800562 (a G-to-A substitution in Chromosome 6, nucleotide position 26092913)
Deletion or insertion polymorphism	DIP/INDEL	One or more nucleotides is either inserted or deleted. Can be as small as one nucleotide long	rs1799752 (the variation is an insertion of 50 nucleotides)
Copy number variation	CNV	A genomic region is repeated end-to-end. May correlate with gene expression levels; most but not all are tagged and represented by SNPs	CNVR2845.14 (esv19484; a large segment on Chromosome 6 may be present in one or more copy)
Short tandem repeats (microsatellites) and variable number tandem repeats (minisatellites)	STR and VNTR	A group of nucleotides is repeated many times and in different numbers of times in different chromosomes. Less common than SNPs; highly polymorphic; more useful in linkage studies than association studies. Also useful for admixture mapping	D6S1276 (a tetranucleotide repeat, TCTA, is repeated a varying number of times in different chromosomes)
Structural variation	—	Having alternative genes or none in a specific region	HLA-DRB region (the *HLA-DRB1* gene is always present on every chromosome, but sometimes one gene—*DRB3*, *DRB4*, or *DRB5*—accompanies it)
Transcriptomic variation	—	Variation in gene expression levels; splicing patterns; interactions with noncoding RNA (ncRNA); epigenetic effect	Any effect on transcription levels of a gene (via transcription factors, ncRNA, or epigenetic modifications) or the transcription product of a gene (alternative splicing)

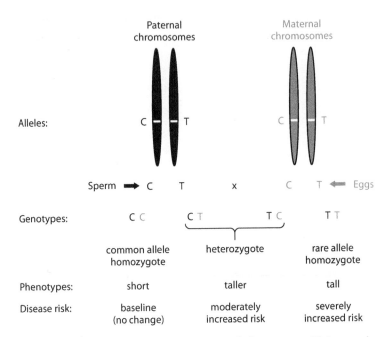

Figure 1.2 Relationships between alleles, genotypes, and phenotypes. Alleles may be any of the four nucleotides (A, C, G, or T). Here, a C (common allele) to T (rare allele) polymorphism is shown. The phenotype example is based on a quantitative phenotype, and disease risk is assumed to correlate with the number of rare alleles in the genotype. For illustrative purposes, both parents are heterozygotes (CT).

of each double-stranded DNA should be used. To distinguish a sequence of two consecutive nucleotides on the same strand from two nucleotides on different strands paired by hydrogen bonds, the sequence of consecutive nucleotides is shown with a "p" between them, meaning that a phosphodiester bond binds them together on the same strand. This nomenclature is most commonly used to denote CpG, which contains the most commonly methylated cytosine nucleotide. CpG is therefore C–phosphodiester bond–G and is different from C:G, indicating a base pair in the same dsDNA strand, or CG, which indicates a genotype consisting of two nucleotides in the same position in the same strand in two chromosomes (see Figure 1.4).

It may be confusing to think that DNA is acidic (deoxyribonucleic acid) despite being made up of the nucleotides that consist of nitrogenous bases. However, the nucleotides also contain deoxyribose and phosphate groups, and it is the phosphate group that makes the DNA an acid.

Nucleotides on the same strand of the DNA form a haplotype

The diploid human genome is made up of two haploid genomes: maternal or paternal chromosomes. A sequence feature in either haploid genome is a **haplotype**. A haplotype is usually a nucleotide sequence on the same chromosome that, in the absence of recombination, is inherited as a unit. An example is the ATG sequence in Chromosome 1p in Figure 1.4. A haplotype may be formed by consecutive nucleotides (like ATG) or nucleotides separated by others (CTAG is another haplotype in Chromosome 1p in Figure 1.4). Sometimes it may even be a single marker representing a haplotype, to mean the chromosome segments of undefined length flanking either side of the marker. In any case, all

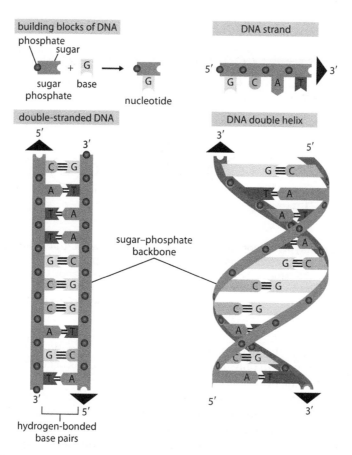

Figure 1.3 DNA structure. DNA is made of four types of nucleotides, which are linked into a DNA strand with a sugar–phosphate backbone from which the bases (A, C, G, and T) extend. A DNA molecule is composed of two DNA strands held together by hydrogen bonds between the paired bases. The arrowheads at the ends of the DNA strands indicate the polarities of the two strands, which run antiparallel to each other in the DNA molecule. In the diagram at the bottom left of the figure, the DNA molecule is shown straightened out; in reality, it is twisted into a double helix, as shown on the right. (From Alberts B, Johnson A, Lewis J et al. [2014] Molecular Biology of the Cell, 6th ed. Garland Science.)

nucleotides or markers making up a haplotype reside on the same chromosome. Typical examples are a C-C-A-T-A haplotype consisting of alleles at five SNPs in a single gene, an *HSPA1B-HSPA1L* haplotype consisting of alleles at these two genes, or an *HLA-DRB1*0401* haplotype consisting of all other alleles usually on the same chromosome with the *HLA-DRB1*0401* allele. Haplotype construction from diploid genotype data requires separation of the alleles on the two strands, a process called **phasing**. This is most easily achieved by using family data when it exists, but there are also statistical methods that attempt to probabilistically separate diploid genotypes into two phased haplotypes.

DNA replication is not error free, and errors generate variation

It is estimated that during each DNA replication event—that is, at cell division—a misincorporation of a nucleotide occurs once in 10^8 to 10^{12} nucleotides. There are mechanisms for proofreading and repair based on the double-helix structure of DNA, but not all errors

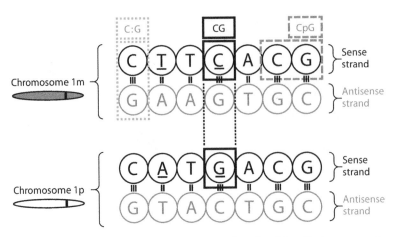

Figure 1.4 Allele, genotype, haplotype, and sequence of nucleotides. Chromosomes 1m and 1p are a homologous pair of chromosomes. The two polymorphic positions (SNPs) are underlined; other nucleotides are the same on both chromosomes on the same strands. C:G is a base pair; CG is a genotype made up of alleles C and G at the same position on the same strands of each homologous chromosome. CpG is two consecutive nucleotides on the same strand of the DNA.

are corrected, leading to the spontaneous appearance of a nucleotide mutation. A spontaneous mutation is initially in only one cell, but when mutations occur in a germ cell, they can be transmitted to the next generation. What happens next is unpredictable. The mutation may be lost at the next generation or may continue to be transmitted to following generations. If possession of this mutation is advantageous to the individuals by increasing their survival and reproductive fitness, the mutation has a higher chance of getting transmitted and gaining a higher frequency in the population. Besides spontaneous replication errors, mutations may also arise due to increased intrinsic metabolic activity or exogenous harmful exposures, which may be physical (for example, ultraviolet radiation) or chemical (for example, aflatoxin or benzene). DNA repair capacity, which may be influenced by genetic factors, plays a role in the ultimate rate of mutation generation.

Nucleotide changes are not random

One might think that when a mutation occurs, any nucleotide may be substituted by any of the others with equal probability. In fact, some substitutions are more common than others, the most common being substitution of cytosine by thymine (shown as C > T). When cytosine loses an amino group, it becomes uracil. This common process, called deamination, may be spontaneous or chemically induced. Uracil is then replaced by thymine in the next replication, and a C-to-T substitution has occurred. Cytosine is also the most commonly methylated nucleotide, especially when it is flanked by a G (CpG), and deamination of methylated cytosine turns it to a thymine, again causing a C > T substitution.

Another common change is caused by oxidative DNA damage which converts guanine (G) to 8-oxy-7,8-dihydrodeoxyguanine. This derivative of guanine (dG) mispairs with adenine (A) instead of cytosine (C), and the original base pair G:C becomes dG:A, which is then converted to T:A at the next replication. Thus, the nucleotide G is now replaced by a T (G > T substitution). Spontaneous deamination of adenine (A) is also possible and results in hypoxanthine (hX), which can pair with cytosine (C). The new pairing of hX:C is corrected to G:C at the next replication, leading to an A > G substitution.

The pairings of A and T, and C with G, are not equal in strength. G and C pairs are stabilized by three hydrogen bonds between them as opposed to the two bonds between A and T (see Figure 1.3). Genomic regions rich in C and G are more difficult to work with due to the strength of physical binding between G and C nucleotides. **DNA methylation**, the most common **epigenetic change**, is the addition of a methyl group to cytosine. Methylated cytosine molecules are denoted as Cm and may be referred to as the fifth nucleotide. Other epigenetics changes are discussed later in the chapter.

Gene expression can be altered with or without changes in DNA sequence

The coding DNA sequences in the genome dictate the primary amino acid sequences of the resulting polypeptides. There are multiple steps from DNA to polypeptide and each step is influenced by multiple genetic and nongenetic factors. Therefore, identical DNA sequences do not necessarily produce identical phenotypes. Even genetically identical monozygotic twins may be different in their disease susceptibility due to differences in their environment. Environmental factors sometimes cause changes in DNA sequence but they more commonly induce biochemical changes without changing the sequence. These changes are called epigenetic changes and may involve DNA itself or the other molecules that package it. The most common end result of epigenetic changes is variation in gene expression, which is also the most common change caused by genetic variants (see Figure 1.1). Sequence variants located in regulatory regions of a gene cause alterations in gene expression. Such regions are not only in or near a gene but may also be present far away from the gene in **intergenic regions**.

Transcription is the conversion of DNA to RNA

The expression of a gene begins with the conversion of the antisense strand of the gene to precursor ribonucleic acid (RNA) inside the nucleus of a cell (**Figure 1.5**). Thus, the RNA made from the antisense strand has the same sequence as the sense strand. Nuclear DNA never leaves the nucleus. RNA is also a nucleic acid but differs from DNA by being a single-stranded molecule and containing the ribonucleotide uracil (U) instead of thymine (T). The process of conversion of DNA to RNA is called **transcription** and involves nuclear **transcription factors (TFs)** acting on their target genes. TFs are gene-encoded proteins that can be seen as major switches that turn on and off gene transcription by binding to their recognition sequences. It is estimated that there are around 2000 TFs in mammalian cells, and most of them work together to enable cells to respond to signals such as hormones, dietary factors, and drugs.

TFs may be highly selective for one or a few genes but are more likely to have very many target genes. For example, once activated, TP53 (the guardian of the genome) can affect many downstream target genes involved in DNA repair and apoptosis. Each TF has a TF binding site (TFBS), which is usually a short nucleotide sequence, located in their target genes. The binding of TF to this site can either promote or repress DNA transcription. Genes usually have multiple binding sequences for different TFs in the regions that regulate their expression. The heaviest concentration of TFBSs is within the **promoter regions** of genes; therefore, sequence variants in this region of a gene have, on average, greater functional effects.

Only a fraction of the human genome codes for peptides, but a large proportion of the genome is transcribed. Most of the transcripts arising from these non-protein-coding, gene-free parts of the genome are noncoding RNA fragments (ncRNA) that have strong

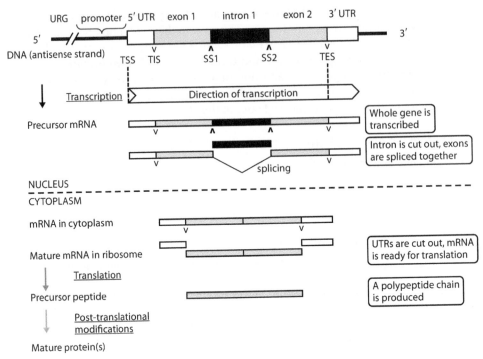

Figure 1.5 Steps in gene expression from transcription to post-translational modifications. First, the whole gene is transcribed to RNA, but introns and untranslated regions (UTRs) are spliced out of messenger RNA (mRNA) before translation. Following the translation of the mature mRNA to a polypeptide, there may be some biochemical changes called post-translational modifications. URG, upstream regulatory region (distal regulatory elements); TSS, transcription start site; TIS, translation initiation site (the first amino acid of the peptide, which is usually methionine, is coded here); TES, translation end site (codes for the last amino acid of the peptide); SS1 and SS2, splice sites 1 and 2, corresponding to acceptor and donor sites.

regulatory roles. With this information, the concept of a gene has started to evolve, and it is now common for the genome region encoding a noncoding RNA to also be called a gene.

RNA undergoes processing

The end result of gene transcription is the primary transcript, which is an RNA strand that will undergo processing. This primary transcript is called heterogeneous nuclear RNA (hnRNA) or precursor messenger RNA (pre-mRNA). The processing takes place in the nucleus and the resulting mRNA leaves the nucleus. The primary transcript is transcribed from the antisense strand so that the resulting pre-mRNA has the same sequence as the sense (or coding) strand, except that thymine (T) is now replaced by uracil (U). The initial transcript undergoes a series of processing steps; these start with **splicing**, where the non-coding sections of the gene sequence are removed, and end with the marking of the beginning and end of the mRNA strand (see Figure 1.5). Splicing of pre-mRNA is a crucial step in the expression of more than 90% of eukaryotic genes that have intervening sequences between coding regions. This process is regulated by fixed sequences called **splice sites** at the beginning and end of introns. Splicing may occur in alternative ways in different tissues or be altered due to polymorphisms affecting splice sites. When a polymorphism abolishes the activity of a splice site, splicing occurs at the next splice site; this leaves out a

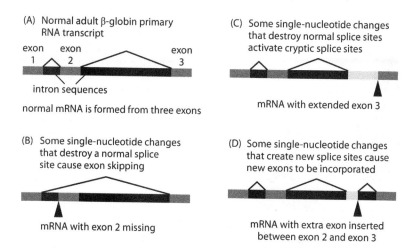

Figure 1.6 Examples of splicing errors in the β-globin gene that cause β thalassemia. (A) Normally, all three exons are included in the mRNA but, due to different mutations (black arrowheads), the mRNA may consist of different exons. (B) The mutation in the splice site at the end of intron 1 causes exon skipping; (C) exon 3 includes the end of intron 2; and (D) a new exon is created from the intron 2 sequence. All of these errors change the structure of the β-hemoglobin, resulting in the disease β thalassemia. (From Alberts B, Johnson A, Lewis J et al. [2014] Molecular Biology of the Cell, 6th ed. Garland Science.)

whole exon and changes the size and nature of the resulting protein. This phenomenon is called **exon skipping** and may generate disease-causing variation (**Figure 1.6**).

Translation is the conversion of mRNA to a polypeptide

The conversion of mature mRNA to a polypeptide chain is called **translation** and takes place in the ribosomes in the cytoplasm (see Figure 1.5). The ribosomes themselves are made up of proteins and RNA molecules called ribosomal RNA (rRNA). The ribosomes are assembled in the nucleolus within the nucleus. Another type of RNA that is involved in the translation process is transfer RNA (tRNA). The tRNA molecules read the code on mRNA—the code for each amino acid consists of three nucleotides (called a codon)—and bring the corresponding amino acid to the ribosome to form the polypeptide chain (**Figure 1.7**). The resulting chain of amino acids that comes out of the ribosome is called a peptide if it is only a few amino acids long, and is otherwise called a polypeptide. For small proteins, this may be the final product, but for most proteins, further modifications occur to produce the mature protein.

The polypeptide undergoes post-translational modification

Further changes to the polypeptide chain are called **post-translational modifications**. Most of these are biochemical changes that take place in the organelle in the cytoplasm called the Golgi apparatus. A common modification is removal of the first few amino acids to convert a propeptide to a peptide (examples include insulin, hepcidin, and endothelin-1). The first amino acid in a propeptide is usually a methionine, but the majority of mature proteins do not start with a methionine. Other possible biochemical modifications include addition of lipids (resulting in lipoproteins), sugar molecules (resulting in glycoproteins), phosphate (phosphorylation), acetate (acetylation), a hydroxyl group (hydroxylation), or

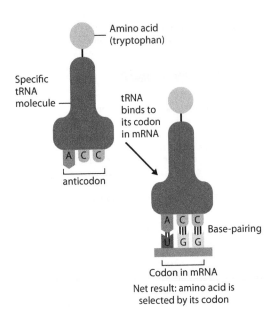

Figure 1.7 Interaction of transfer RNA with messenger RNA in the translation process. The tRNA molecule is specific for tryptophan and has the anticodon (ACC) that recognizes the codon for tryptophan (UGG) on the mRNA molecule. (From Alberts B, Johnson A, Lewis J et al. [2014] Molecular Biology of the Cell, 6th ed. Garland Science.)

a methyl group (methylation), or making biochemical bridges between amino acids of the same or different polypeptides. One such bond is a disulfide bridge that occurs between the thiol groups of cysteine molecules of polypeptides (**Figure 1.8**). Any variation that changes the cysteine to another amino acid would result in the loss of a disulfide bond in the protein structure if that cysteine is involved in such a bridge.

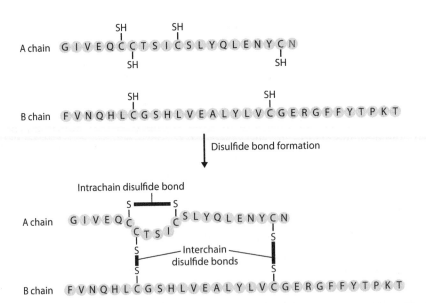

Figure 1.8 Disulfide bonds in the insulin molecule. Disulfide bonds form between the sulfhydryl (SH) groups on the side chains of cysteine residues. They can form between cysteine side chains within the same polypeptide (such as between positions 6 and 11 within the insulin A chain) and also between cysteine side chains on different, interacting polypeptides such as the insulin A and B chains. (From Strachan T & Read A [2010] Human Molecular Genetics, 4th ed. Garland Science.)

From thousands of genes to millions of protein products

Proteins are usually modified versions of the original polypeptides and commonly consist of more than one polypeptide chain joined together. The estimated number of protein-coding genes in the human genome is around 20,000, but the proteome consists of more than one million different proteins. The increase in the number of products from a smaller number of genes starts at the gene level. It is possible to have several different mRNA products from a single gene sequence due to a variety of processes such as alternative promoter usage, **alternative splicing**, or mRNA editing. The main source of variety is, however, post-translational modification, which may vary with time or location for the same polypeptide (**Figure 1.9**). Furthermore, the different folding of proteins consisting of the same polypeptide(s) may add yet more variety. With there being fifty times more protein products than the number of genes in the human genome, it is naive to expect that, by studying the gene sequence variation, a direct correlation can always be made with a phenotype. This is one of the reasons why even the largest and most sophisticated genetic association studies cannot on their own explain all of the **heritability** of a phenotype.

Gene expression is highly regulated

Gene expression is a highly regulated and selective process. All genes are not expressed at all times or in all cells—gene expression shows specificity to cell types and varies across time. There is a very strong influence of the environment on gene expression. Genetic association studies should collect as much information on environmental exposures as possible to be able to analyze joint effects of genes and environment. It is also possible that some genetic variants are simply modifiers of environmental effects, and, without inclusion of the environmental factors, a purely genetic study may be unable to detect an association. This is the basis of **effect modification** or **gene and environment interaction (GxE)**.

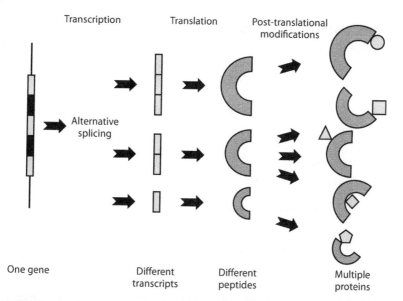

Figure 1.9 **Multiple proteins can be generated from one static gene sequence, mainly by alternative splicing and post-translational modifications.**

The results of the landmark Encyclopedia of DNA Elements (**ENCODE**) project substantially changed our view on genome biology. Until the completion of this project, the prevailing view was that only a fraction of the genome was coding for proteins and the majority of the genome had no function (it was even called junk DNA). It used to be believed that the most crucial polymorphisms were those in coding regions that cause changes in amino acid sequence, followed in importance by promoter-region polymorphisms that change the sequence of TFBSs. It is now better appreciated that the vast spaces between genes also contain very important information relevant to gene expression regulation. These intergenic regions encode a variety of noncoding RNA sequences with a multitude of functions in the regulation of transcriptional and translational machinery. As a result, more than 85% of associations reported with **genome-wide statistical significance** are SNPs located in intergenic regions. It has also been shown that the association of intergenic-region SNPs with disease is also due to their correlation with gene expression variation.

Most disease-associated SNPs are expression quantitative trait loci

Gene expression is a quantitative trait that is measured on a continuous scale, usually by measuring gene-specific mRNA or, sometimes, protein levels. Any genetic polymorphism that shows a correlation with mRNA or protein levels is called an **expression quantitative trait locus** or **eQTL**. Until recently, promoter-region SNPs that alter TFBS sequences, plus a few other SNPs in promoter-like regions called enhancer or silencer regions, were the best-known eQTLs. This view is now changing with the realization of the function of intergenic regions. SNPs in regions that do not code for proteins are frequently within transcribed noncoding RNA sequences and influence expression levels of genes that may even be in different chromosomes. If a SNP correlates with the expression of a nearby (usually within 5 Mb) gene, it is called a *cis*-**eQTL**. A SNP that correlates with expression levels of genes far away in the same chromosome, or even in a different chromosome, is a *trans*-**eQTL**. Selecting SNPs that are eQTLs is now a widely used strategy to increase the statistical power of detecting an association.

The correlation between gene and protein expression levels is not exact

Most gene expression studies examine mRNA levels as a proxy for the overall expression of protein-coding genes. The final product of gene expression, however, is the protein. There are multiple steps from transcription of the DNA to production of the mature protein where there may be interference with the process and modification of the levels and function of the ultimate protein product. Known determinants of the expression of a protein product include the following:

- The rate of transcription (modified by genetic variants and environmental factors)
- hnRNA processing to yield mRNA
- The transport of mRNA to the cytoplasm
- The stability of the mRNA, which determines the rate of its degradation
- The rate of translation of mRNA to form a polypeptide
- Post-translational modification of the peptide product
- The rate of transfer of the protein to its target (for example, to the cell surface)
- The stability of the protein, which determines the rate of its degradation

The ideal determinant of gene expression is therefore direct measurement of relevant tissue-specific protein levels and function, but this is logistically and technically highly demanding except for proteins within easily collected body fluids (for example, blood). The correlation between mRNA levels and protein levels has been extensively examined, and the best estimates suggest no more than 50% correlation. The discrepancy is believed to be due to the modifications involving noncoding RNA species. Gene expression studies based on mRNA measurements should therefore be interpreted with caution. Furthermore, given the variation in the expression of genes depending on cell types and environmental exposures, studies examining mRNA levels in single cell types at resting state may not reveal any effect of genetic variations on gene expression. In such studies, measurement of gene expression in the cell types of interest following induction of expression by the correct environmental factor (such as heat shock, hormones, or cytokines, for example) should yield more information. The **Genotype-Tissue Expression (GTEx)** program in the National Institutes of Health (NIH) aims to examine the correlations between genome-wide DNA sequence variants and mRNA levels measured by the most sensitive method in multiple tissues for better interpretation of eQTL effects. The results are publicly accessible via the GTEx portal.

1.2 The Human Genome

The human genome is made up of 3 billion nucleotides (**Box 1.1**). Only 1–2% of the genome makes up the coding regions of around 20,000 protein-coding genes. These genes are located in 22 pairs of autosomal chromosomes and the two sex chromosomes (X or Y chromosomes). Each chromosome exists in two copies; one inherited from the father (paternal copy) and one inherited from the mother (maternal copy). Pairs of each chromosome are called **homologous chromosomes**. Due to this pairing of chromosomes, the human genome is a diploid genome that results from the union of the haploid genomes of two germ cells in sexual reproduction. In the case of sex chromosomes, females have two copies of homologous X chromosomes, while males have one X and one Y chromosome. This is an important distinction and crucial in the analysis of sex chromosome variation. Although each chromosome in a homologous pair is generally similar to the other, this is not always the case. The same gene may or may not exist in the homologous member, or may exist in a different copy number (**Figure 1.10**). Most genes, however, exist as one copy per haploid genome and therefore are called single-copy genes.

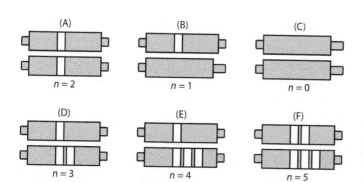

(A)   (B)   (C)

$n = 2$ $n = 1$ $n = 0$

(D)   (E)   (F)

$n = 3$ $n = 4$ $n = 5$

Figure 1.10 Different types of copy number variation. (A) Normally, a gene exists as a single copy in each homologous chromosome; (B) heterozygote deletion; (C) homozygote deletion; (D, E, F) each homologous chromosome has at least one copy of a gene. n denotes the total number of copies of the gene in the diploid genome.

The reference sequence provides standard coordinates for each gene and polymorphism

With thousands of genes and millions of SNPs, the human genome's content is very rich and it is important to have reliable information about the location of genes and other genome features. In the past, different laboratories devised their own systems to detail locations of genomic features, but these had to be standardized in a consensus system. Such a system is now in place and the full **reference sequence (RefSeq)** of each chromosome is used to mark all features of that chromosome, especially variants at each nucleotide position.

Box 1.1 The human genome in numbers

There are 3 billion nucleotides in nuclear DNA (compared with 5 million nucleotides in *E. coli* and 16,500 nucleotides in the mitochondrial genome). The largest genome of any organism belongs to *Paris japonica* (Japanese canopy plant), which is 50 times larger than the human genome.

There are 23 pairs of chromosomes (compared with 5 in *Drosophila* and >250 in some ferns).

Chromosomes vary in size from 50 Mb (Chromosome 21) to 263 Mb (Chromosome 1); 98.5% of the human genome does not code for proteins, but 80% is transcribed.

If the entire genome was protein coding, there would be 1.3 million protein-coding genes.

There are more than 20 million common single nucleotide polymorphisms; each personal genome differs from the reference sequence in ~3.5 million common SNPs; each human has around 3 million SNPs in a heterozygous state, of which 12,000 are missense variants.

The proportion of nucleotides that differ between human and chimpanzee genomes is 1.3%.

The proportion of nucleotides that differ between any two human genomes is 0.1%.

Approximately 5% of the human genome is conserved in comparison to mammalian genomes, due to regulatory roles common to mammalian genomes.

There are 20,000 protein-coding genes (compared with <5000 in *Drosophila* and approximately 19,000 in *Caenorhabditis elegans*).

There are more than 25,000 noncoding RNA genes.

There are more than 14,000 pseudogenes (defunct relatives of functional genes) that are not translated to proteins but may be transcribed.

The average gene size is 10,000 to 15,000 nucleotides; in nucleotide numbers, the largest gene is the dystrophin gene (*DMD*), which is 2.4 million nucleotides long and encodes a protein 3685 amino acids long (in comparison, insulin is 51 amino acids long and is encoded by a gene 1430 nucleotides long); in terms of coding region or amino acid length, the titin gene (*TTN*) encodes the largest protein (34,350 amino acids); the shortest protein-coding gene (<500 nucleotides) encodes a histone (but the shortest RNA-coding genes are transfer RNA genes at 70 to 80 nucleotides).

The mitochondrial genome also contains genes and exists in 500 to 1000 copies in the cytoplasm of the cell (as opposed to, on average, two copies of each gene in the nucleus).

The DNA is split among the chromosomes; if it were a single, continuous thread, it would be almost two meters long.

More than 85% of associations are with SNPs in noncoding regions.

The average number of exons in a gene is 8.8, resulting in an average coding-region size of 1340 nucleotides.

The location of any genomic feature is first identified by the chromosome it belongs to. By convention, the short arm of a chromosome is called the p arm and the long arm is the q arm. Human chromosomes have a centromere separating the long and short arms. The nucleotide numbering of each chromosome begins from the tip (telomere) of the short arm and ends at the tip of the long arm. RefSeq provides coordinates, or nucleotide positions, for each variant and gene. For example, the total length of Chromosome 6 (short and long arms together) is 170 Mb (million bases); the *HFE* gene is located in Chromosome 6p between nucleotide positions 26,087,281 and 26,096,117. Each SNP is given a unique identification number in the reference sequence: the *HFE* SNP rs1800562 is a G > A variation at nucleotide position 26,092,913.

RefSeq does not come from a single individual. It is a consensus sequence and the sequence of different chromosomes, or even different parts of the same chromosome, may be from different individuals. Having a consensus sequence is not ideal, as the same chromosomes are not the same size in different people and so the position of features will not be in exactly the same place, but RefSeq provides a useful framework to work with. The reference positions of each gene and polymorphism can be found in the National Center for Biotechnology Information (NCBI) databases Entrez Gene and Entrez SNP, respectively.

Genomes also contain pseudogenes

As well as amino acid-coding genes, genomes also contain **pseudogenes** (see **Box 1.1**). They are either remnants of extinct genes or nonfunctional copies of existing genes. Although they no longer encode functional polypeptides, some may still be transcribed to RNA fragments and act as regulators of gene expression by **RNA interference**. The most relevant aspect of the existence of pseudogenes in genetic association studies is that they are a nuisance factor, their presence being a major obstacle in the design of certain genotyping assays, plus they interfere with successful amplification of the intended fragments of functional genes. Regardless of their expression status, pseudogenes or functional genes that have sequence similarities to other genes in the human genome are collectively called **paralogs**.

There are differences between nuclear and cytoplasmic DNA

The overwhelming majority of DNA in a cell is contained within the cell nucleus, packaged into chromosomes, and is known as **nuclear DNA**. Total nuclear DNA is contributed to virtually equally by both parents. Cells also have DNA outside the nucleus, called extranuclear or cytoplasmic DNA, but as it is located in mitochondria, it is usually referred to as **mitochondrial DNA (mtDNA)**. Mitochondria are cytoplasmic organelles where cellular respiration occurs and mtDNA encodes some of the enzymes required for the biochemical reactions of respiration. The mitochondrial genome is much smaller than the nuclear genome (16.5 kilobases or kb versus 3,000,000 kb) but is present in much greater copy numbers: 500 to 1000 copies in each cell. The abundance of mtDNA is the reason for its use in most fossil studies, since more mtDNA than nuclear DNA is still present in fossil cells. The two types of DNA in the cell do not react or mix with each other. Another significant difference between them is that mtDNA is inherited only from the mother, and thus use of mtDNA in ancestry determination is more accurate than use of nuclear DNA.

DNA is packaged by histones

DNA does not exist as a single linear stretch, even within each chromosome. It is packed in a highly elaborate system of coiling or twisting, which has implications for regulation of

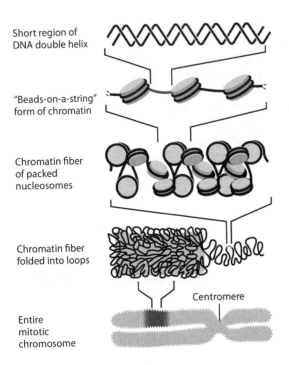

Short region of
DNA double helix

"Beads-on-a-string"
form of chromatin

Chromatin fiber
of packed
nucleosomes

Chromatin fiber
folded into loops

Centromere

Entire
mitotic
chromosome

Figure 1.11 The complex structure of DNA in a chromosome. DNA is first wrapped around histone proteins to form nucleosomes, which make up chromatin when further packaged. Chromatin then forms the chromosomes. It is the epigenetic changes on the DNA (methylation) or histone proteins (more than 60 biochemical changes including methylation) that regulate gene expression via modification of the access of regulatory proteins to DNA. (From Alberts B, Johnson A, Lewis J et al. [2014] Molecular Biology of the Cell, 6th ed. Garland Science.)

gene expression. Small proteins called **histones** form cores around which DNA is wound in a double loop. Around each disc-shaped histone core (consisting of eight histone molecules) are wrapped 146 base pairs of DNA and this makes the DNA look like beads on a string (**Figure 1.11**). These beads are called nucleosomes. The total package of DNA and histone proteins is called **chromatin**, the coiling of which forms the chromosomes. Access of DNA-binding proteins to their targets on DNA is regulated by this packaging, either by the chemical state of histones or the location of the DNA in relation to the histone proteins. Histones may undergo more than 60 biochemical or epigenetic changes that regulate access to promoters, or such regulatory regions may be located between nucleosomes and thereby remain freely accessible. These accessible sites are called hypersensitivity sites and are where transcription is most active because there is no interference from histone molecules. The most common types of post-translational modification of histone proteins are acetylation, methylation, phosphorylation, and ubiquitylation. These modifications have different effects on histone molecules and subsequently on gene expression regulation. For example, the methylation of lysine 4 on histone 3 is correlated with active gene expression.

There is even more to discover

With the sequencing of 1000 human genomes in the **1000 Genomes Project**, it has become apparent that in each genome sequenced there are megabases of DNA sequence of unknown nature that cannot be mapped to the reference sequence. These unknown sequences are enriched for repeated elements but also contain functional elements. Thus, the Human Genome Project can be seen as still ongoing. There are still some unknowns about the human genome even at the most simple level of DNA sequence, and work continues to unravel the mysteries of genome biology.

1.3 Consequences of Genetic Variation

Genetic variation may modify protein structure

While alterations of gene expression levels are the most common intermediate mechanism of observed disease associations with DNA variants (see Figure 1.1), there are other ways for causal variants to show associations with traits. If the variation is in the coding region, it may be a silent variant that does not change an amino acid. This is possible because there is redundancy in the genetic code, and most amino acids are coded by more than one nucleotide triplet (codon). Variants that change the triplet but still code for the same amino acid are called **synonymous variants** (synonymous SNP; sSNP). **Nonsynonymous variants** (nsSNPs) cause amino acid substitutions (**Figure 1.12**). This substitution can be either a single amino acid change (**missense variant**) or, if the **coding sequence** is shifted due to a deletion or insertion of a nucleotide, the rest of the amino acid sequence is affected (**frameshift**). The most severe change is the introduction of a premature stop codon, which will truncate the expressed protein (**nonsense variants**). Large screening studies have concluded that most missense variants that are rare in populations have deleterious effects and are the most promising candidates to show strong associations with disease susceptibility. The most extreme examples of rare missense variants are single mutations causing monogenic disorders such as cystic fibrosis or phenylketonuria.

Sequence changes may also have serious consequences for protein structure due to alteration of interactions between amino acids—for example, the loss of disulfide bonds due to a missense variant substituting another amino acid for cysteine. Specific examples of such variants are given later in this chapter.

Genetic variants can cause favorable or unfavorable functional changes

There is no direct correlation between the type of variant and its outcome: a SNP in a crucial gene may result in dire outcomes, but loss of a whole gene that belongs to a physiologic

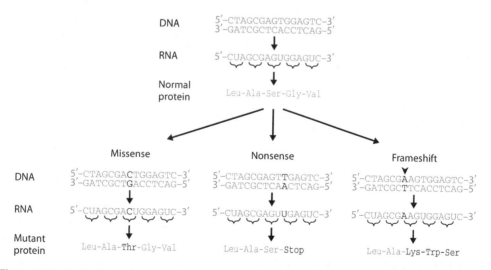

Figure 1.12 Nucleotide substitutions in coding regions may cause amino acid sequence changes. These could be an amino acid substitution (missense: Ser is substituted by Thr), introduction of a stop codon (nonsense: GGA is changed to TGA), or insertion of a nucleotide that changes the amino acid sequence from that point onward (frameshift).

pathway with lots of redundancy may be tolerable. As an example, a single nucleotide substitution in the tumor suppressor genes *TP53* or *BRCA1*, if present in the germ line, may cause inherited cancer. On the other hand, the 32-nucleotide deletion in the *CCR5* gene known as the Δ32 deletion does not cause any functional defect, since CCR5 is a member of a large protein family that consists of proteins with similar functions. Besides not having a deleterious effect, the *CCR5* Δ32 deletion has a protective effect against infection with human immune deficiency virus (HIV) through a nongenetic mechanism. It is the location within the gene and which gene it affects that is more important in terms of the variant's phenotypic effects. The *CCR5* Δ32 deletion is also a reminder that variants are not exclusively harmful; some may have beneficial effects. The direction of the change in gene function (such as decreased or increased expression) and the context (whether higher expression is beneficial or not in a tissue) determine whether the variation will be a risk or protective marker.

The location of the variant determines the outcome

To better appreciate the correlation of sequence variations with functional consequences, it is helpful to consider gene structure and how gene activity is regulated. For a variant to have an effect on gene function, it has to interfere with the expression of the gene or change the molecular structure of the protein product (**Figure 1.13**). Each gene consists of a promoter region, which is where the signal for transcription of DNA to RNA is initiated; **exons** or coding sequences (CDS) encoding amino acids of the subsequent protein; **introns** or intervening sequences (IVS) between exons; and finally a sequence encoding a **poly-A tail**. The presence of introns in eukaryotic genes is common but intronless or single-exon genes also exist. It is estimated that around 3% of human genes are intronless.

Introns allow diversification of the protein pool by alternative splicing, which allows the generation of multiple polypeptides from the same gene. About half of human genes are subject to alternative splicing. Alternative splicing is regulated by splice-site sequences within introns, so sequence variants affecting these sites may cause alterations in the

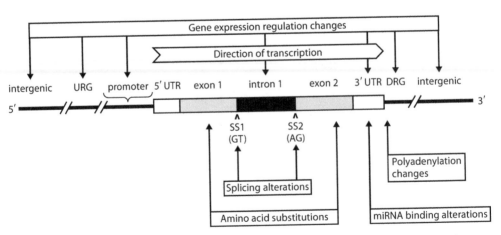

Figure 1.13 Structure of a typical gene and important sites in the regulation of gene activity or protein structure. DNA sequence variants in these sites are more likely to have functional roles and to show associations with traits. URG, upstream regulatory region (distal regulatory elements); UTR, untranslated region; SS1/SS2, splice sites 1 and 2, corresponding to acceptor and donor sites; DRG, downstream regulatory region.

splicing of gene products. Alternatively spliced forms of proteins (encoded by the same gene) may have different functions, which may in turn affect disease susceptibility. One example is the peroxisome proliferator-activated receptor-γ (*PPARG*) gene and its isoforms, type 1 and 2. The two isoforms are formed by alternative splicing, which is caused by SNP rs1801282. The type 2 isoform is associated with reduced risk for type 2 diabetes and the type 1 isoform is associated with increased risk.

The promoter (at the 5′ end of a gene) contains some general TFBSs and other sites that influence the tissue-specific expression of the gene. TFBSs may also be present in other parts of the gene and also in intergenic regions. The 5′ end of the first exon to be transcribed and the 3′ end of the last exon to be transcribed contain sequences that are transcribed to RNA but not translated to protein. These regions are called the **5′** and **3′ untranslated regions (UTRs)**, respectively (see Figure 1.13). The 3′ UTR may have **microRNA**-binding sites that regulate gene expression. While TFs initiate expression of a gene, microRNAs usually suppress the expression levels post-transcriptionally. Another important aspect of the 3′ UTR is the presence of a **polyadenylation signal** (a hexanucleotide sequence, AATAAA), which is responsible for the poly-A tail of approximately 200 adenosine nucleotides at the end of gene transcripts. Variants in the polyadenylation signal can cause increased disease risk; for example, a polymorphism in the polyadenylation signal of *N*-acetyltransferase 1 (*NAT1*) causes higher acetylation activity and is associated with increased risk for multiple cancers. Most but not all genes encode poly-A tails. Most notably, histone gene transcripts do not have poly-A tails. The 3′ UTR is also involved in mRNA stability, transport of mRNA out of the nucleus, and efficiency of translation.

Polymorphisms altering the function of any of these gene structures are called **functional** or **pathogenic variants** and may show disease associations. Most polymorphisms, however, lack any effect on gene activity and are referred to as neutral, silent, or nonfunctional polymorphisms. Naturally, these variants do not usually show associations with any trait, unless they are correlated with functional variants nearby. **Table 1.2** lists common functional variants and specific examples known to modify gene activity.

Functional variants also occur outside genes

Genes are aligned on chromosomes and generally have some space, known as intergenic regions, between them. Similar to the intervening sequences (introns) between exons in a gene, these intergenic regions do not have amino acid-encoding sequences, but may contain sequences important in regulating gene activity. Such sequences can be TFBSs or response elements, microRNA, or microRNA-binding sites. Intergenic regions may also contain additional promoter elements called distal promoters. Sequence variation in these elements has functional consequences (see Table 1.2). Furthermore, such regulatory elements may translocate to other parts of the genome during rearrangements that accompany cancer development and cause aberrant expression of genes in their new location.

The variants included in an association study are not necessarily assessed for functionality a priori, but this assessment is done after an association is shown. In one of the most educational experiences in genetic epidemiology, a group of SNPs in an intergenic region (the q24 region of Chromosome 8) showed strong associations with multiple cancers in different studies. These markers are in the middle of a 1.18 Mb-long region that does not contain any genes. Once these puzzling results were obtained, the function of the markers was evaluated. After a series of elegant experiments, the SNPs in this gene desert turned out to influence the transcription of the nearest gene—the *MYC*

Table 1.2 Examples of functional variants and associated traits

Variant type	Variant	Disease or trait, and mechanism
Promoter	*TNF* rs1800629	Breast cancer, autoimmune disorders
	HMOX1 STR	Type 2 diabetes, cancer via transcriptional regulation
Intronic	*LTA* IVS1 SNP rs909253	Myocardial infarction via transcriptional regulation
	FGFR2 IVS2 rs2981578	Breast cancer via altered TF binding
	IRF4 IVS4 SNP rs12203592	Childhood leukemia via transcriptional regulation
	COL1A1 IVS1 SNP rs1800012	Osteoporosis via altered TF binding
Exonic	*BTNL2* rs2076530	Sarcoidosis via alternative splicing leading to a truncated protein
	PPARG rs1801282	Type 2 diabetes via alternative splicing
	TNFRSF1A rs1800693	Multiple sclerosis via alternative splicing
	LDLR rs688	Increased cholesterol levels via alternative splicing leading to exon skipping
	HFE rs1800562 (C282Y)	Hereditary hemochromatosis via amino acid substitution abolishing a disulfide bond
	ABCB1 (*MDR1*) exon 26 synonymous SNP rs1045642	Cancer via altered substrate specificity
	DRD2 synonymous SNP rs6277	Dopamine-induced up-regulation of *DRD2* expression via decreased mRNA stability and translation
3′ UTR	*HLA-C* rs67384697	HIV-1 progression via microRNA-binding site change
	HLA-G rs1704 (14-base-pair indel)	Recurrent pregnancy loss and pre-eclampsia via alteration of mRNA stability
	PARP1 rs8679	Bladder and breast cancer via microRNA-binding site change
Complete gene deletion	*C4A* deletion	Systemic lupus erythematosus
	CYP21A2 deletion	Congenital adrenal hyperplasia
	GSTM1 and *GSTT1* deletion	Cancer
	MICA deletion	Nasopharyngeal cancer
Intergenic	Multiple chromosome 8q24 SNPs	Breast, prostate, and colon cancer via long-range interaction with *MYC* in a tissue-specific manner

proto-oncogene—as very-long-distance regulators. This example shows that even when it is not obvious, variants may have functional consequences.

Most functional polymorphisms are associated with traits due to the changes they cause in gene expression levels

Sometimes it is easy to predict that a variant is functional; for example, when it is in the promoter region of a gene and the nucleotide substitution affects the interaction between a transcription factor and its binding site. It is estimated that around one-third of polymorphisms in a promoter region are of functional significance, and the likelihood generally increases as the distance to the beginning of the gene decreases. Thus, the nearer a polymorphism is to the **transcription initiation site**, the more likely that it will affect gene transcription levels. Functionality of a polymorphism can be assessed either computation-ally (*in silico* methods) in a "dry" laboratory or by various transcription assays in a "wet" laboratory. More directly, correlations between polymorphisms and gene expression levels can be examined in population studies or in cell lines. The discipline dealing with these types of studies is known as **genetical genomics** and has contributed to the assessment of polymorphisms. For most polymorphisms that show associations with traits, functional replication data may already be available in published papers, usually in large supplementary data files, or in online databases that compile such data for easy access.

Variants in coding regions are likely to be functional

Nucleotide substitutions in coding regions may result in amino acid substitutions, which may influence the protein structure in many ways. One example is the substitution of cys-teine with tyrosine in amino acid position 282 (C282Y) in the hereditary hemochromato-sis protein encoded by *HFE*. The substitution of tyrosine in place of cysteine at position 282 prevents the formation of a disulfide bond between two cysteine molecules, and thus the whole protein structure changes. As a consequence, the HFE molecule loses the ability to attach itself to the cell membrane and cannot function. The end result is increased iron absorption leading to the iron overload disease hemochromatosis.

Synonymous or silent coding-region SNPs are not always nonfunctional. They may still alter the function of an exonic transcriptional enhancer. A rare example of an exonic silent SNP altering the function of a gene is given for the multi-drug resistance gene *MDR1*. In this case, the variant influences the dynamics of translation and this change creates a functional effect.

Total deletion of genes is possible

The most severe change in gene activity that is possible is its total loss, which can result from deletion of the gene. This is not as uncommon as one might think, though due to the redundancy in the human genome, and mechanisms such as **canalization**, not all dele-tions are clinically significant. Complete gene deletions are common for complement component genes *C4A/C4B*, and *GSTM1* and *GSTT1*, and have some functional effects (see Table 1.2). Total deletion of *CYP21A2* is rare but abolishes 21-hydroxylase enzyme activity and causes congenital adrenal hyperplasia if it occurs in two copies or if there is another deleterious mutation on the second copy of the gene.

Partial gene deletions are also possible. In the majority of French Canadians with familial hypercholesterolemia, a large portion of the gene for the low-density lipoprotein receptor (*LDLR*) is deleted. Most crucially, the deletion covers the entire promoter, leading to no expression of the gene. Total loss of gene activity can also result from a nonsense variant. The *C4A* gene may be physically deleted, but may also be rendered nonfunctional by a two-nucleotide insertion in exon 29. This insertion introduces a stop codon and no functional protein is produced.

Copy number variation has functional consequences

Complete deletion of a gene is a form of CNV, but an increase in gene copy number is more common than deletion. Having the same gene in multiple copies may not affect its activity, but there are examples in which increased copy number shows direct correlation with gene expression levels (as assessed by mRNA levels in cell culture). One example is the correlation between a CNV of the *FCGR3A* gene and the expression and function of its product, the protein FcγRIIIa, in natural killer (NK) cells, where it regulates the production of the human chemokine CCL3-L1. Most CNVs have correlations with nearby sequence variants. For example, the CNV associated with rheumatoid arthritis susceptibility (CNVR2845.14) is marked by the SNP rs2395185. Current genome-wide association microarray chips include millions of SNPs as well as most CNVs for genotyping, and coverage can be assumed to be complete.

Functional change may require some interaction

An individual polymorphism may not show an association with a trait but may still be very important in determination of that trait. This is possible if a polymorphism interacts with another polymorphism. In this case, it is possible that each polymorphism shows no association individually. This is a gene-by-gene interaction, also called **epistasis**, in the absence of a detectable **main effect**. Another possibility is that an individual polymorphism may still be a crucial determinant of the phenotype while showing no individual main effect. A typical example is the tissue types determined by human leukocyte antigen (HLA) genes. HLA genes are the most polymorphic expressed genes. HLA polymorphism at the protein level has substantial functional consequences, but is determined by multiple amino acid changes, and each amino acid change may be determined by more than one polymorphism. In this case, single nucleotide changes in the DNA sequence may appear to be neutral, but constellations of them defining the HLA type may altogether yield a strong association. This issue has been addressed in data analysis by using algorithms.

1.4 Genetic Nomenclature

There is an almost unlimited number of possible changes that can occur in a genome and the use of standard nomenclature to define them is a challenge. Although there are efforts on many fronts to standardize the reporting of genetic variation, the results of these efforts are materializing only slowly. It is not uncommon to see published papers referring to variants with traditional names rather than the unique identification (ID) numbers provided by NCBI Entrez SNP. Each and every sequence variation in the human genome now has a reference sequence ID beginning with rs and followed by a unique ID number: for example, rs1800562 defines the SNP that causes the amino acid substitution C282Y in *HFE*. Originally, most SNPs were named in relation to the nucleotide position and the nucleotide substitution. An example is the tumor necrosis factor (TNF) promoter region SNP, −308 G/A, which has shown associations with many diseases. It is possible that even the most recent studies may use the customary rather than the standard ID for this SNP, which is rs1800629. The conversion from customary variant names to standard names is made possible by the HuGE Navigator Variant Name Mapper, and thus there is no excuse not to use the standard nomenclature.

The most accessible guidelines for sequence variant nomenclature are provided online by the Human Genome Variation Society. Some of the most common rules for gene and variant naming are as follows:

- All human genes are shown in capital letters and in italics (*HFE* not HFE or Hfe).
- The protein products are still shown in capital letters but not in italics (*HFE* gene and HFE protein).
- Human gene names do not contain a hyphen, with the exception of *HLA* genes (*IL6* not *IL-6*; *HLA-B* is fine).

- SNPs should be described by their Entrez SNP standard IDs. However, if the customary names are given, the nucleotide substitution should be shown as major allele > minor allele. Thus, the TNF SNP rs1800629 corresponds to −308 G > A (but not G/A or G − 308A). The number in customary names denotes the position of the substituted nucleotide in relation to the translation initiation site (beginning of the gene). The minus sign denotes that the position of the nucleotide is not downstream but upstream of the beginning of the gene. In this example, −308 corresponds to a position within the promoter of the gene. If coding-region SNPs are cited with their amino acid positions and changes, then the common amino acid and the variant amino acid are shown using the one-letter amino acid codes on either side of the position number. The correct application of this rule to the *HFE* variant is C282Y, but not 282 C/Y or 282 C > Y.

Key Points

- Gene expression is the transcription of a DNA sequence to an mRNA sequence, followed by translation of the mRNA sequence to an amino acid sequence, which becomes polypeptides and proteins. The whole process is highly regulated.

- Gene expression can be assessed by measuring mRNA levels or protein levels. The correlation between mRNA and protein levels is not exact.

- There are 20,000–22,000 protein-coding genes, which account for about 1.5% of the genome. The rest of the genome (intergenic regions) contains sequences that are important for regulation of gene expression.

- Some intergenic sequences are transcribed into noncoding RNA, which acts to regulate gene expression.

- Genetic variation is the key ingredient of association studies. Whether such variants represent the total deletion of a gene or a single nucleotide variation, they are important when they alter the function of the gene or its product.

- Coding-region variants usually modify disease susceptibility by causing changes in protein structure, but may also be involved in regulation of splicing or gene expression.

- Variants with the highest level of functionality are those located within regulatory elements such as promoters, enhancers, and silencers. This is the most common form of phenotypic modulation by a genetic variant.

- Functional variants may be far away from the genes they modify and may be within intergenic regions.

- DNA sequence changes are not the only type of variants relevant in genetic association studies; there are also structural and copy number changes.

URL List

dbSNP (Database of single nucleotide polymorphisms). National Center for Biotechnology Information. http://www.ncbi.nlm.nih.gov/snp

ENCODE Explorer. Nature. http://www.nature.com/encode

ENCODE Project. National Human Genome Research Institute. http://www.genome.gov/10005107

Gene. National Center for Biotechnology Information. http://www.ncbi.nlm.nih.gov/gene

GENCODE Human Genome Statistics. http://
www.gencodegenes.org/stats.html

GTEx (Gene and Tissue Expression) Portal.
Broad Institute. http://www.gtexportal.org

HuGE Navigator Variant Name Mapper. http://
www.hugenavigator.net/HuGENavigator/
home.do

NCBI Entrez. National Center for Biotechnology
Information. http://www.ncbi.nlm.nih.gov/
gquery

Nomenclature for the description of
sequence variants. Human Genome
Variation Society. http://www.hgvs.org/
mutnomen

Further Reading

Introduction to molecular genetics

Al-Chalabi A (2009) Genetics of Complex
Human Diseases. A Laboratory Manual. Cold
Spring Harbor Laboratory Press.

Attia J, Ioannidis JP, Thakkinstian A et al. (2009)
How to use an article about genetic association:
A: Background concepts. *JAMA* 301, 74–81 (doi:
10.1001/jama.2008.901).

Feero WG, Guttmacher AE & Collins FS (2010)
Genomic medicine – an updated primer.
N Engl J Med 362, 2001–2011 (doi: 10.1056/
NEJMra0907175).

Guttmacher AE & Collins FS (2002) Genomic
medicine – a primer. *N Engl J Med* 347, 1512–
1520 (doi: 10.1056/NEJMra012240).

Knight JC (2009) Genetics and the general
physician: insights, applications and future
challenges. *QJM* 102, 757–772 (doi: 10.1093/
qjmed/hcp115).

Strachan T & Read A (2010) Human Molecular
Genetics, 4th ed. Garland Science. (*Covers the
topics included in this primer in more detail
as well as more advanced topics in molecular
genetics, with a focus on human genetics. The
second edition of this book is accessible via the
NCBI Bookshelf; http://www.ncbi.nlm.nih.gov/
books/NBK7580.*)

Tefferi A, Wieben ED, Dewald GW et al.
(2002) Primer on medical genomics part
II: Background principles and methods in
molecular genetics. *Mayo Clin Proc* 77, 785–808
(doi: 10.4065/77.8.785).

Functional polymorphisms

Cooper DN (2010) Functional intronic
polymorphisms: Buried treasure awaiting
discovery within our genes. *Hum Genomics* 4,
284–288 (doi: 10.1186/1479-7364-4-5-284).

Guo Y & Jamison DC (2005) The distribution
of SNPs in human gene regulatory regions.
BMC Genomics 6, 140 (doi: 10.1186/1471-2164-
6-140).

Knight JC (2005) Regulatory polymorphisms
underlying complex disease traits. *J Mol Med
(Berl)* 83, 97–109 (doi: 10.1007/s00109-004-
0603-7).

Maston GA, Evans SK & Green MR (2006)
Transcriptional regulatory elements in
the human genome. *Annu Rev Genomics
Hum Genet* 7, 29–59 (doi: 10.1146/annurev.
genom.7.080505.115623).

Ng PC & Henikoff S (2002) Accounting for
human polymorphisms predicted to affect
protein function. *Genome Res* 12, 436–446
(doi: 10.1101/gr.212802).

Wang X, Tomso DJ, Liu X & Bell DA
(2005) Single nucleotide polymorphism
in transcriptional regulatory regions
and expression of environmentally
responsive genes. *Toxicol Appl Pharmacol*
207(2 Suppl), 84–90 (doi: 10.1016/j.
taap.2004.09.024).

Copy number variation

Conrad DF, Pinto D, Redon R et al. (2010)
Origins and functional impact of copy
number variation in the human genome.
Nature 464, 704–712 (doi: 10.1038/
nature08516).

Gamazon ER, Nicolae DL & Cox NJ
(2011) A study of CNVs as trait-associated
polymorphisms and as expression quantitative
trait loci. *PLoS Genet* 7, e1001292 (doi: 10.1371/
journal.pgen.1001292).

Primer on Medical Genetics

2

Medical genetics traditionally deals with clinical conditions purely or largely due to genetic causes such as single-gene (Mendelian) disorders. More than 4500 Mendelian disorders are known but they constitute only 1% of all diseases. Information used by medical geneticists to identify these rare diseases has strong implications for genetic association study design, execution, and interpretation. Most current genetic association studies are typically for **multifactorial** or **complex disorders**, which are determined by the interplay among multiple gene variants and environmental factors, but the principles of medical genetics are important in understanding the nature of multifactorial disorders and setting up informative studies. It should be noted that mutations causing Mendelian disorders were located using **linkage studies**, which are not covered in this book. This chapter will review the main principles of medical genetics and how they connect to the study of multifactorial disorders by association studies.

2.1 The Location of Disease-causing Mutations

Genes with disease-causing mutations can be in any part of the human genome

Disease susceptibility is modified by a mutation that changes the activity of a gene. The terms "mutation" and "polymorphism" are frequently used interchangeably, but the differences are listed in **Table 2.1**. Every polymorphism is a mutation, but it is only referred to as a polymorphism when it occurs in more than 1% of the population. In the context of disease risk, most mutations cause disease in most of their bearers (they have high **penetrance**), but polymorphisms only show a weak effect on predisposition to disease. While disease-causing mutations are usually missense coding-region variants, most polymorphisms change the expression level of a gene; this can be a gene in the vicinity (*cis*-effect; which is usually defined as within 5 Mb), or at a distance (>5 Mb away), or even on a

Table 2.1 Main differences between mutations and polymorphisms

	Mutation	Polymorphism
Frequency	Rare (≤1%)	Common (>1%)
Role in disease development	Causal; necessary and sufficient (deterministic)	Weak modifying/predisposing effect; neither necessary nor sufficient
Effect size (relative risk)	Very high (>10)	Small to modest (1.1 to 3.0)
Penetrance	Very high (~100%)	Low (~1–2%)
Inheritance pattern	Clear (Mendelian)	Equivocal
Identification	Linkage studies in families	Association studies in unrelated individuals

different chromosome (***trans*-effect**). The gene whose function is changed by a variant is called the **disease gene** if the change is strong enough to be instrumental in disease pathogenesis. The affected gene may be on an autosomal chromosome and the disease it causes is evident in both males and females (autosomal genetic diseases), or it can be on a sex chromosome, in which case one sex is more likely to be affected than the other (sex-linked genetic diseases). The **mitochondrial genome** also contains genes whose variants can cause serious medical conditions.

The majority of disease genes carrying disease-causing mutations are on autosomal chromosomes

Given that 22 of the 23 pairs of chromosomes are autosomal, a great majority of mutations occur in genes on these chromosomes. However, the effects of autosomal genetic variants may be influenced by either the products of genes on a sex chromosome or sex hormones, and may therefore show sex specificity. Examples of sex-specific traits include male pattern baldness and hairy ears, which are encoded by genes in autosomal chromosomes but expressed only in males. This point has to be remembered in genetic association studies: differences in the incidence rates of diseases between the sexes do not necessarily mean the disease gene is on a sex chromosome. Sometimes, the effect of the variant on an autosomal gene depends on the parental origin of the chromosome. This is called a **parental effect** and is discussed in a later section of this chapter.

Only three of the 23 pairs of chromosomes (Chromosomes 21, 13, and 18) can be found in a trisomic state compatible with life. The reason for this is that these chromosomes are the ones with the fewest genes and they can be tolerated in a trisomic state. Fetuses with multiple copies of larger and more gene-dense chromosomes are lost before birth. There is nothing special about the genes on Chromosomes 21, 13, and 18 that cause certain syndromes when present in three copies. Likewise, disease genes are equally distributed in autosomal chromosomes and there is generally no preferential distribution of them on any particular chromosome. Possible exceptions are the high number of disease associations with HLA-region variants and the disproportionately high number of immune system-related genes in the X chromosome.

Disease-causing mutations are also present in genes located on sex chromosomes

The human sex chromosomes are the X and Y chromosomes. The Y chromosome contains around 100 unique protein- and non-protein-coding genes that have no counterparts on the X chromosome, but there are no inherited disorders due to a mutation in a disease gene on the Y chromosome. Microdeletion of the azoospermia factor (AZF) region of the Y chromosome is common in oligospermia/azoospermia, and there are some suggestive studies for associations of Y chromosome variants with cardiovascular disease risk. It should be noted that the Y chromosome is about three times smaller than the X chromosome and a portion at both tips is shared by both chromosomes (**Figure 2.1**). These shared portions are called **pseudoautosomal regions** (PAR1 and PAR2) and act as autosomal genes. Approximately 95% of the Y chromosome is unique and is called the male-specific region. An important feature of this region is that it does not recombine, due to the lack of a corresponding chromosome with which to exchange material. This means that the Y chromosome is transmitted as invariant blocks from male to male. These blocks, consisting of specific variants, are called haplogroups (**Figure 2.2**) and are widely used in

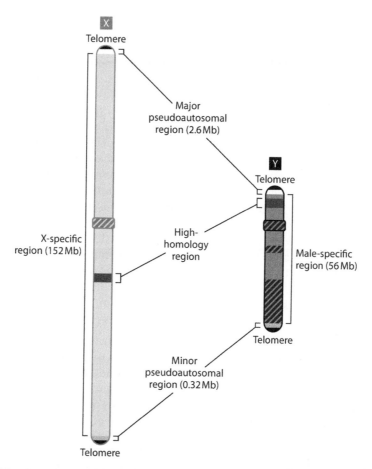

Figure 2.1 Shared pseudoautosomal regions at the tips of X and Y chromosomes. These regions behave like autosomal genes rather than as sex chromosomes in inheritance patterns. (From Strachan T, Goodship J & Chinnery P [2015] Genetics and Genomics in Medicine. Garland Science.)

phylogenetic studies. Genetic association studies examining why there may be sex disparity in certain diseases have recently started to use the Y chromosome haplogroups. The microarray chips used in genome-wide association studies have only recently included SNPs from the Y chromosome and more data on Y chromosome polymorphisms and disease susceptibility should become available soon. There is no reason to exclude these

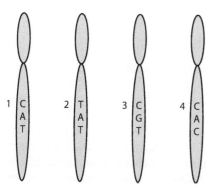

Figure 2.2 Haplogroups are formed by different constellations of multiple polymorphisms on the same chromosome. In this example, three SNPs (C > T; A > G; T > C) form multiple haplotypes: haplogroup 1 is represented by C-A-T; haplogroup 2 by T-A-T; haplogroup 3 by C-G-T; and haplogroup 4 by C-A-C.

SNPs from any study. However, data from the Y chromosome should be analyzed separately as allelic associations and, obviously, only in males.

While any Y chromosome-linked trait is transmitted from male to male, traits encoded in the X chromosome can be transmitted from either parent but are usually expressed in males. The preferential expression in males is due to the presence of only one copy of the X chromosome in males (females have two copies). While both males and females express dominant mutations, which require only one copy, females need to have two copies of a recessive mutation to express the trait. Males express single-copy recessive mutations because there is no second copy of the X chromosome to mask the effect of the recessive mutation. The imbalance in the number of X chromosomes in males and females creates some difficulty in the statistical analysis of data from X chromosome SNPs. Early association studies either did not include the X chromosome in the study or did not analyze the data. With the advent of statistical methods that make these analyses feasible, X chromosome associations are now being examined. Such analysis should not be restricted to disorders with sex effects because any SNP on the X chromosome, especially in PARs, may be involved in any disease with or without a known sex-specific presentation.

Mitochondrial disease genes can only be inherited from the mother

While Y chromosomes are only transmitted to the offspring from the father, the mitochondrial genome is transmitted only by the mother (to both male and female offspring). Depending on the cell type, each cell contains from a few to hundreds of mitochondria, and each mitochondrion has several copies of the mitochondrial genome. The result is that each cell contains an average of 500 copies of the mitochondrial genome. Mitochondrial DNA predominantly encodes proteins that are components of the mitochondrial respiratory chain. Some nuclear genes influence mitochondrial functions, so the source of a mitochondrial disorder may not exclusively be in the mitochondrial genome. Due to the variable number of copies of the mitochondrial genome in each cell, cells may contain a mixture of mutant and wild-type mitochondrial genomes, a situation called **heteroplasmy**, which may lead to phenotypic variability among the offspring of the same mother. The ratio of wild-type and mutant chromosomes may also shift during each cell replication. Because redox reactions take place in mitochondria, the production of reactive oxygen species is high and the *de novo* **mutation** rate is around 1000 times higher in mitochondrial DNA compared with nuclear DNA. This results in sporadic disorders caused by mitochondrial DNA mutations. The high mutation rate also generates a lot of polymorphisms that are useful markers for disease association studies and, in the case of mitochondrial haplogroups, for phylogenetic studies. Comprehensive genetic association studies should include mitochondrial DNA SNPs in their genotyping schemes.

2.2 Single-Gene, Oligogenic, and Multigenic Disorders

Medical geneticists traditionally deal with genetically well-characterized conditions called **single-gene disorders**, where a gene becomes nonfunctional through a high-penetrance mutation. Most of the well-documented medical genetic conditions—such as cystic fibrosis, phenylketonuria, congenital adrenal hyperplasia, and rare inherited forms of common diseases such as Parkinson's disease and breast cancer—are single-gene disorders. They are called single-gene disorders but they do not have to be single-mutation disorders as the disease gene is commonly incapacitated by different mutations. Inheritance of these diseases follows Mendelian patterns (**Table 2.2**). Some diseases, however,

Table 2.2 Main features of Mendelian inheritance patterns

	Autosomal dominant	Autosomal recessive	X-linked dominant	X-linked recessive	Y-linked	Mitochondrial
Example	Huntington's disease	Cystic fibrosis	Vitamin D-resistant rickets	Hemophilia	Male infertility	Leber's hereditary optic neuropathy
Multiple generations affected?	YES	**NO** (skips generation)	YES	**NO** (skips generation)	YES	YES
Is a parent always affected?	YES (unless it is a new mutation)	NO	YES	NO	YES	YES
Are both sexes affected?	YES	YES	YES (F > M)	YES (M > F)	**NO**[a]	YES
Are all affected individuals male?	NO	NO	NO	**YES**[b]	**YES**[a]	NO
Male-to-male transmission?	POSSIBLE	POSSIBLE	**IMPOSSIBLE**[c]	**IMPOSSIBLE**[c]	YES	IMPOSSIBLE
Is it always transmitted from the mother?	NO	NO	NO	NO	NO	**YES**[d]
Is it always transmitted from the father?	NO	NO	NO	NO	**YES**	NO
	50% of the children of an affected parent will be affected. Each affected child has an affected parent	One in four children of healthy (carrier) parents will be affected. An affected child usually has unaffected parents	All female children of an affected father are affected	No male children of an affected father are affected	Only males are affected	Transmitted only from an affected mother (no transmission from an affected father)

[a]A father's Y chromosome is always and only inherited by sons (never by daughters); [b]a son receives an X chromosome from his mother and does not inherit his father's X chromosome; [c]a father's X chromosome is always inherited only by a daughter; [d]mitochondrial DNA is only transmitted from the mother.

are caused or modulated by more than one gene. It is now clear that even single-gene disorders are strongly modulated by additional (modifier) genes. Diseases caused by interactions of mutations in several genes are called **oligogenic disorders**. One example is the digenic form of retinitis pigmentosa, which results from simultaneous heterozygous mutations in two genes, *ROM1* and *RDS*. Another disease that has a digenic form is Hirschsprung disease, where the *RET* and *GDNF* genes are mutated. In Bardet-Biedel syndrome, three mutations in two genes are needed for disease development (tri-allelic inheritance). **Multigenic disorders** are more common than single-gene or oligogenic disorders and are caused by mutations or functional variants in many genes. The genetic basis of multigenic disorders is explored by most current genetic association studies. Even in single-gene disorders the environment also plays a role, interacting with genetic variants. For example, phenylketonuria cannot result from the gene mutation only; if the environment (dietary phenylalanine intake) is controlled, the patient will never develop the disease. Thus, almost all diseases can now be called multifactorial.

Different mutations can cause the same phenotype

With increasing knowledge of the human genome, our view of the genetic basis of even the simplest genetic disorders is evolving rapidly. Most single-gene disorders are first attributed to a single mutation. Additional mutations are then recognized either in the same gene and with similarly strong effects on the phenotype or in additional genes that result in a similar phenotype. These two situations are known as **allelic heterogeneity** and **genetic heterogeneity**, respectively (**Figure 2.3**). Genetic heterogeneity is also known as **locus heterogeneity**. These concepts are highly relevant in complex disease genetics as they are always caused by multiple genetic variants. Recently, epigenetic changes that modify gene activity have been added to the mix of genomic changes that cause or modify single-gene disorders.

One condition with tremendous heterogeneity is retinitis pigmentosa. It has monogenic and digenic forms, and a large proportion of cases present with an unknown genetic basis. Another feature of retinitis pigmentosa is that even a certain gene mutation may be inherited via different modes, causing an autosomal recessive inheritance pattern in one family and an autosomal dominant pattern in another. Allelic and genetic heterogeneity both have important implications in genetic association studies. In two different studies, it is not uncommon to find associations with different variants of the same gene. Likewise, failing to replicate an association finding in a second study may be due to genetic heterogeneity and the predominant effect of a different gene in a different population due to different modifiers. Therefore, even negative results may have a plausible biological explanation in light of allelic and genetic heterogeneity.

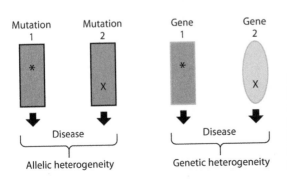

Figure 2.3 Allelic and genetic heterogeneity. In allelic heterogeneity, different variants of the same gene increase the risk for the same disease. In genetic heterogeneity, variants in different genes independently increase the risk for the same disease.

A combination of multiple genetic and nongenetic factors leads to multifactorial inheritance

As opposed to single-gene disorders, disorders such as diabetes, rheumatoid arthritis, schizophrenia, and most cancers are determined by a large, but unknown, number of genetic and nongenetic variables, making them multifactorial traits. The genetic variables participating in susceptibility to multifactorial diseases are generally low-penetrance variants that would yield very small, if any, **effect sizes** on their own. It is their interaction with one another that generates the **genetic load** for increased susceptibility. Susceptibility to a multifactorial disease can be seen as a quantitative trait, and when the accumulation of genetic and environmental factors exceeds a critical threshold, the disease occurs. This concept is known as **Falconer's polygenic threshold model** for dichotomous non-Mendelian characters (**Figure 2.4**). There are several implications of this model:

- The greater the number of predisposing risk genes possessed by the parents, the greater the probability that they will have an affected offspring, depending on the contribution of the environmental factors.

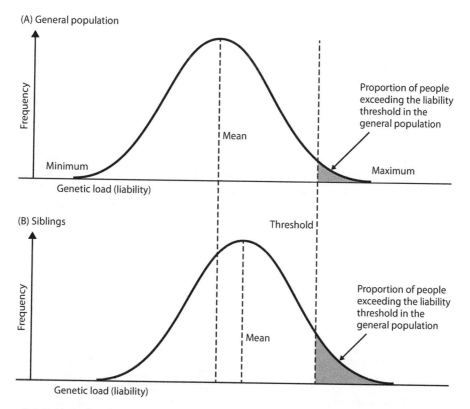

Figure 2.4 Falconer's polygenic threshold model. (A) The distribution of liability or the amount of genetic contributors to risk in the general population shows a normal distribution, with most people having average liability. Only those whose liability exceeds the polygenic threshold develop the disease. (B) Siblings of diseased individuals have higher liability on average, and a higher proportion of siblings have a liability exceeding the threshold. Thus, a higher proportion of siblings develop the disease than the proportion of the general population.

- The disease may show a familial aggregation but without a discernible Mendelian segregation pattern.
- Familial risk declines with increasingly remote degrees of relationship (from first-degree relatives toward third- and lower-degree relatives).
- The greater the number of affected family members, the higher the recurrence risk in other family members.
- Recurrence risk increases with severity of the disorder in the index case.
- When there is a marked difference in incidence between sexes, it is because of differences in risk threshold between males and females. The less frequently affected sex has a higher risk threshold and, simply because of greater load, transmits the condition more often to the more frequently affected sex, which has a lower threshold.

Single-gene disorders are now being considered more as multifactorial disorders with one major mutation and multiple modifiers. The distinction between single-gene disorders and multifactorial or complex disorders is therefore becoming blurred. However, in a complex trait, there is no main determinant and the interaction of multiple modifier genes results in the disease. The main implication for genetic association studies relates to study design and the need to have comprehensive coverage of genetic and environmental variables.

2.3 Copy Number of Mutant Genes

The term "disease gene" is often used, but what is meant by this is a gene that is rendered less functional by mutation. Deactivation of a gene may result from both copies being deleted or inactivated due to deleterious mutations (or epigenetic changes), or one copy being inactivated and the remaining copy being either insufficient to maintain normal function (haploinsufficiency) or rendered inactive by the mutant copy (dominant-negative mutation), usually by physical interaction (**Figure 2.5**). Mutations on one or both copies of a gene can therefore cause disease. The nature of the mutation and whether one or two copies of it cause the disease determines the inheritance pattern (dominant or recessive).

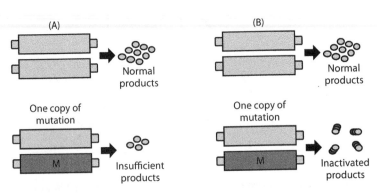

Figure 2.5 Two mechanisms for how a single copy of a mutation renders the gene inactive. (A) In haploinsufficiency, one intact copy of the gene does not produce a sufficient amount of the protein product. (B) A dominant-negative mutation results in the production of a mutant peptide that inactivates the normal product. In both instances, not enough normal product is available to carry out the function.

The location of the gene further determines the autosomal or sex chromosome inheritance pattern. If there is a discernible pattern in disease transmission for a complex disease, determining this pattern may prove useful in the design and analysis of genetic association studies. The main features of inheritance patterns (including mitochondrial transmission) are summarized in Table 2.2. In a complex disease, the mode of action of an individual SNP is usually unknown. It may therefore be necessary to examine different genetic risk models for different inheritance patterns for SNPs in genetic association studies rather than assuming that one model is powerful enough to yield associations for any model.

Disease-causing mutations are propagated by a variety of mechanisms

It is counterintuitive that a mutation that kills its bearers, even before their reproductive age, may continue to be present in the population. It is, however, possible for mutations, especially those that cause recessive diseases, to propagate themselves mainly via unaffected heterozygotes. Although those individuals who develop the disease may not be able to pass the mutant gene to the next generation, there will be a few individuals who have not yet developed the disease or who have developed a milder form and they will transmit the mutation. Furthermore, it has been documented for a number of recessive disease-causing mutations that they may even confer an advantage to the carriers of one copy of the mutation (in heterozygous form). This phenomenon is one form of **heterozygote advantage** and most commonly occurs by conferring resistance to an infectious agent; such mutations remain in the population. Some examples of advantages conferred by disease-causing mutations are shown in **Table 2.3**. For any homozygous offspring who has inherited two copies of the mutant gene, there are two heterozygous siblings who have inherited only one copy, and those individuals will

Table 2.3 Suggested mechanisms for the maintenance of deleterious mutations in the population

Disease	Mutant gene	Advantage for heterozygotes
Sickle-cell anemia	*HBB*	Resistance to *Plasmodium falciparum*
α-Thalassemia	*HBA1/2*	Resistance to *Plasmodium falciparum*
Hemoglobin C	*HBB*	Resistance to *Plasmodium falciparum*
Glucose 6-phosphate dehydrogenase deficiency	*G6PD*	Resistance to *Plasmodium falciparum*
Cystic fibrosis	*CFTR*	Resistance to cholera toxin and *Salmonella enterica* serovar Typhi; increased fertility
Congenital adrenal hyperplasia	*CYP21A2*	Increased fertility; brisk cortisol response in stress
Hereditary hemochromatosis	*HFE*	Protection from iron deficiency; protection from *Salmonella enterica* serovar Typhimurium infection in mice
Tay-Sachs disease	*HEXA*	Resistance to *Mycobacterium tuberculosis*
Phenylketonuria	*PAH*	Protection from spontaneous miscarriage (via reversal of toxicity of ochratoxin A)
Huntington's disease	*HTT*	Increased fertility, decreased risk of cancer
α₁-Antitrypsin deficiency	*SERPINA1*	Increased fertility and twinning

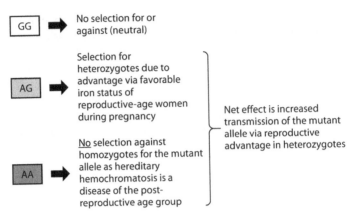

Figure 2.6 The *HFE* mutation C282Y that causes the autosomal recessive disease hereditary hemochromatosis is maintained in the population via heterozygote advantage. The C282Y mutation is due to a nucleotide substitution, G to A. The common genotype GG confers no advantage or disadvantage to its carriers. Heterozygotes (AG) are at an advantage due to increased iron absorption. Because the disease occurs later in life, selection against mutation homozygotes (AA) does not occur. Selection for heterozygous women of reproductive age maintains the frequency of the mutation in the population.

continue transmitting the disease mutation. An example of propagation of a disease-causing mutation is given in **Figure 2.6**.

These observations have implications for genetic association studies. Heterozygote advantage is real and should be considered in the analysis phase of a genetic association study. Heterozygote advantage is also highly relevant in infectious disease susceptibility determined by HLA antigens. The observations listed in Table 2.3 are relevant to SNP selection or prioritization in genetic association studies. If there are indications that a SNP or a genomic region has been subject to selective pressures, those SNPs should be given priority over others since this is evidence for their functionality.

Heterozygosity for mutations causing recessive diseases may have functional effects

Recessive traits are caused by mutations in both copies of a gene, although the mutations in each copy may be different, a situation known as **compound heterozygosity**. Hereditary hemochromatosis shows compound heterozygosity for *HFE* mutations C282Y and H63D, and congenital adrenal hyperplasia shows compound heterozygosity for a number of *CYP21A2* mutations. While such allelic heterogeneity is well recognized, there has been less interest in the effects of heterozygosity for recessive mutations. Heterozygosity does not cause the disease—or even a milder form of clinically detectable disease—in carriers, but, at least for biochemical traits, there is evidence that biochemical changes are detectable. For example, in heterozygotes for *HFE* mutations, serum iron parameters are often mildly changed but do not come close to the iron overload levels seen in some homozygotes. Likewise, heterozygotes for *CYP21A2* mutations have a lesser degree of the biochemical signs of hormonal changes that are seen in homozygotes with full-blown disease. Although there is considerable overlap of the biochemical values observed in heterozygotes and mutation-free healthy people, a proportion of heterozygotes have hormone levels slightly above normal values. Such people are called **biochemically manifesting heterozygotes** or just manifesting carriers. Heterozygosity for mutations causing rare recessive disorders is not as rare as homozygosity. In certain populations, up to 50% of individuals may be heterozygous for *HFE* mutations. It is estimated that in Europeans, carrier frequencies for cystic fibrosis and factor V Leiden deficiency are 1/29 and 1/14, respectively, but there has been no long-term follow-up of the carriers of recessive disease mutations for health outcomes. Recessive mutations should not be dismissed when they exist in one copy and should be included in genetic association studies whenever they are relevant to the phenotype under study.

Polymorphisms in disease genes may also modify disease risk

Disease-causing mutations are generally highly functional mutations that cause a severe loss of gene function. The *HFE* C282Y mutation, for example, causes a total loss of the cell surface expression of HFE protein. *CYP21A2* mutations similarly decrease the enzymatic activity of the gene product to zero or near zero. The tumor suppressor gene *TP53* may be totally inactivated by its cancer-causing mutations. However, disease-causing mutations are not the only changes in the respective genes. Like all genes, genes involved in single-gene disorders also have low-penetrance polymorphisms, mainly in the form of SNPs, and some of these will be functional. The effects on gene function of low-penetrance SNPs are much less than those of disease-causing, high-penetrance mutations, but they may be detectable in a genetic association study. For *HFE*, *CYP21A2*, *TP53*, *BRCA1*, and *BRCA2*, such functional polymorphisms have been identified and associations with various traits have been reported. Thus, polymorphisms of disease genes may still contribute to disease susceptibility even when there are no mutations, and they should be considered in genetic association studies when relevant.

2.4 Inherited Mutations

The mutations that cause inherited disorders are present in germ-line DNA in either, or both, the sperm or egg, so that the offspring inherits the mutation from either or both parents and the mutation is present in all cells of the offspring derived from the zygote. Polymorphisms that modify disease risk are also present in the germ line. Rarely, a mutation may occur during intrauterine development and it will then only be present in the lineage of the mutant cell. This is known as **mosaicism** and describes the co-presence of mutant and nonmutant cells in the body. A well-known example of mosaicism is the Dalmatian dogs' coat, which contains a mixture of mutant and nonmutant pigment cells that generate the typical appearance of spots. Mosaicism is also possible in the germ-cell population of either parent. Mosaicism is frequently sought either in parents or in offspring when an inheritance pattern is difficult to explain. One example is the appearance of an autosomal dominant disorder in multiple offspring when both parents are unaffected. This can happen if either one of the parents has germ-cell mosaicism where the mutation is present in their germ cells but not in their somatic cells. The situation could also arise if all affected offspring had *de novo* mutations that occurred during their development, but this is very unlikely. A mosaic parent with a mild form of a single-gene disorder can have offspring that are more severely affected. In this case, the offspring will have the mutant gene in every cell of their body while the parent only has the mutation in a proportion of their somatic cells.

Undetected mosaicism is probably more common than detected mosaicism but does not cause great concern in a genetic association study. In other words, one should expect to find the same genotype in any biospecimen collected from the same individual.

Therefore, it is acceptable, but not ideal, to have DNA extracted from peripheral blood cells in patients and from buccal cells in healthy controls, as is frequently the case in studies of childhood disorders. In a survey of 100 families with children with genetic disorders, 4% of the healthy parents were found to have indications of somatic mosaicism. These individuals with somatic mosaicism have a mixture of mutant and nonmutant cells.

Certain environmental exposures cause mutations only in a certain tissue, for example in the skin following ultraviolet (UV) radiation exposure and in liver cells after aflatoxin exposure. These mutations may cause local effects (skin cancer or liver cancer in these examples) but would not be detectable in other tissues such as peripheral blood cells. Cancer-related mutations detected in genomic studies of tumor cells, such as the ongoing The Cancer Genome Atlas (TCGA) study by the NIH, do not exist in other somatic cells of affected individuals. The implications of these findings in genomic association studies are not clear but are probably minimal. Likewise, such somatic mutations may not have strong implications for understanding what has caused that cell to turn into a cancer cell. What is important in genetic association studies is to identify the variations that predispose individuals (hosts) to cancer development; these are likely to be very different from those detected in an end-stage tumor cell, which has gone through multiple rounds of genomic changes including structural, mutational, and epigenetic alterations. Thus, designing a genetic association study in cancer to examine the variants of a gene that has been found to be highly expressed in a tumor cell, but not in a normal host cell, would not have a strong justification.

Mutations do not always produce equal effects in populations

Two populations with the same frequency of a disease-causing mutation may have different frequencies of the disease, and diseased individuals in each population may have different average levels of disease severity. These two situations are determined by the penetrance and **expressivity** of the mutation. If only a proportion of people with the mutation develop the disease, it is called reduced penetrance. If all affected people do not suffer from the disease at the same severity, it is called variable expressivity (**Figure 2.7**). Reduced penetrance may even be evident in a single family, for example when an individual with the mutation of an autosomal dominant disorder such as retinoblastoma does not have the disease despite having a diseased parent and a diseased child. This is a case of reduced penetrance in an individual who has the mutation but is missing a crucial contributor to disease development. In addition to genetic and allelic heterogeneity (see Figure 2.3), there are often unknown genetic and environmental modifiers that determine whether a mutation causes a disease and the severity of the disease caused.

The iron overload disease hereditary hemochromatosis, caused by the *HFE* C282Y mutation, is a well-known example of an inherited disease that is strongly modified by other genetic variants and lifestyle factors. Less than 5% of people homozygous for the C282Y mutation develop the clinical disease (although more may have biochemical changes) and not all of them have a severe form of disease. Both the penetrance and expressivity of the C282Y mutation are modified by other factors such as genetic variants of iron regulatory genes that are also influential on iron homeostasis, behavioral factors like alcohol consumption, or regular blood loss via menses or blood donation. Age is also a modifier since the iron accumulation gets stronger with age. The subset of breast cancer caused by inherited *BRCA1* or *BRCA2* mutations is another example of reduced penetrance. These mutations increase the lifetime risk for breast cancer considerably, up to eight times, but not to 100%. Genetic modifiers of these mutations are being identified.

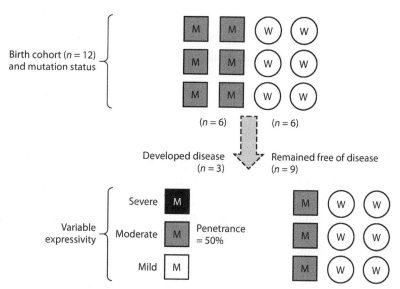

Figure 2.7 Schematic explanation of penetrance and expressivity. M denotes individuals positive for the mutation and W denotes those without the mutation (wild-type genotype). Half of the mutation-positive individuals (3 of 6) developed the disease (thus, penetrance is 50%) with variable severity (expressivity), and all those with the wild-type genotype remained free of disease.

Both penetrance and expressivity (see Figure 2.7) have implications for the design and interpretation of genetic association studies. Since even single-gene diseases can be modified by factors other than the disease mutation, it should be expected that a similar situation exists for genetic variants that modify disease predisposition. It is therefore crucial to collect as much data as possible for genetic and environmental variables in order to have reliable information to help with data analysis and interpretation of results. Such data will be very useful at the analysis stage to identify confounding factors as well as statistical interactions that identify effect modifiers. The current common practice of analyzing SNP data as an individual data point is likely to miss associations that would have been unraveled if the SNP data were analyzed in conjunction with other variables.

Another phenomenon well known to medical geneticists is the occurrence of a disease at younger ages in successive generations. Trinucleotide (triplet) repeat disorders such as Huntington's disease, spinocerebellar ataxia, myotonic dystrophy, and fragile X syndrome are characterized by **genetic anticipation**. With an increasing number of repeats at each generation, called expansion, the age at disease onset gets younger. For some diseases, larger expansions in the trinucleotide repeats are caused by paternal transmission, but for other diseases it is maternal transmission. Such a parental effect is not unique to this group of disorders and is known to exist in other inherited and complex disorders. In diseases either caused or modified by DNA sequence variants, where heterozygosity is sufficient to increase the disease risk, the parental origin of the variant may have to be taken into account to unravel an association. Another situation where a parental effect is observed is in a multifactorial disease such as congenital pyloric stenosis. While the disease is more common in male newborns, children of affected females (the less frequently affected sex) are more likely to be born with the pyloric stenosis than the children of affected males. This is because females need a higher genetic load to express the disease themselves and transmit more of that load to the offspring. Among the offspring of an affected female, males are more likely to express the disease because of the requirement for a lower genetic load. The end result is that the less frequently affected sex transmits the disease more easily to the more frequently affected sex. While family-based association study designs may be able to unravel parental effects, in a **population-based association study** such effects cannot be examined, resulting in a lower chance of identifying an association.

Key Points

- Even in single-gene disorders there are strong modifiers that can be genetic or environmental. This is why, in a genetic association study of a complex multifactorial disease, a single variant will yield a very small effect size.

- A disease-causing mutation is rare and highly deleterious for gene function. A disease-associated polymorphism (like a SNP) is usually common and alters gene function but is not very deleterious.

- Disease-causing mutations can be on autosomal or sex chromosomes, or in the mitochondrial genome.

- Mutations for diseases that show higher frequency in one sex do not have to be in sex chromosomes.

- A Mendelian pattern of inheritance is a feature of very rare diseases; the lack of such a pattern does not rule out a genetic basis for disease.

- Common human diseases are multifactorial, result from complex interactions between genetic and environmental factors, and do not show a straightforward inheritance pattern.

- Heterozygosity for a mutation causing a recessive disease is generally tolerable, and may confer some advantage to its carriers. This is the reason for the maintenance of some disease-causing mutations in the population.

- Even if a variant is strongly associated with disease susceptibility, the mere presence of it is not generally sufficient to determine with certainty that the associated disease will develop.

- Diseases with a strong genetic basis are still modified heavily by other factors, and genetic association studies should therefore be as comprehensive as possible to unravel the joint effects of multiple factors.

URL List

Genetics for Epidemiologists: Application of Human Genomics to Population Sciences. National Human Genome Research Institute. http://www.genome.gov/27026645

Strachan T & Read AP (1991) Human Molecular Genetics, 2nd ed. Bookshelf. National Center for Biotechnology Information. http://www.ncbi.nlm.nih.gov/books/NBK7580

Further Reading

Disease genes (autosomal and sex chromosomes)

Bodmer W & Bonilla C (2008) Common and rare variants in multifactorial susceptibility to common diseases. *Nat Genet* 40, 695–701 (doi: 10.1038/ng.f.136).

Charchar FJ, Bloomer LD, Barnes TA et al. (2012) Inheritance of coronary artery disease in men: an analysis of the role of the Y chromosome. *Lancet* 379, 915–922 (doi: 10.1016/S0140-6736(11)61453-0).

Clayton DG (2009) Sex chromosomes and genetic association studies. *Genome Med* 1, 110 (doi: 10.1186/gm110).

Jimenez-Sanchez G, Childs B & Valle D (2001) Human disease genes. *Nature* 409, 853–855 (doi: 10.1038/35057050).

Inheritance patterns (monogenic versus polygenic)

Badano JL & Katsanis N (2002) Beyond Mendel: an evolving view of human genetic disease transmission. *Nat Rev Genet* 3, 779–789 (doi: 10.1038/nrg910).

Dipple KM & McCabe ER (2000) Modifier genes convert "simple" Mendelian disorders to complex traits. *Mol Genet Metab* 71, 43–50 (doi: 10.1006/mgme.2000.3052).

Wang DY, Chan WM, Tam PO et al. (2005) Gene mutations in retinitis pigmentosa and their clinical implications. *Clin Chim Acta* 351, 5–16 (doi: 10.1016/j.cccn.2004.08.004).

Homozygosity versus heterozygosity

Rotter JI & Diamond JM (1987) What maintains the frequencies of human genetic diseases? *Nature* 329, 289–290 (doi: 10.1038/329289a0).

Sidransky E (2006) Heterozygosity for a Mendelian disorder as a risk factor for complex disease. *Clin Genet* 70, 275–282 (doi: 10.1111/j.1399-0004.2006.00688.x).

Germ-line mutations

Rochette J, Le Gac G, Lassoued K et al. (2010) Factors influencing disease phenotype and penetrance in *HFE* haemochromatosis. *Hum Genet* 128, 233–248 (doi: 10.1007/s00439-010-0852-1).

Stein CM & Elston RC (2009) Finding genes underlying human disease. *Clin Genet* 75, 101–106 (doi: 10.1111/j.1399-0004.2008.01083.x).

Primer on Population Genetics

3

Population genetics deals with changes in genetic variation in a population over time, and how these changes occur. A mutation originally occurs on a single chromosome of an individual and may never be passed on to the next generation. Population genetics answers questions that arise if the mutation is transmitted. Does it reach a detectable frequency in the population? How is that frequency maintained? Is it subject to selection pressure? Studying genetic variations and how they evolve requires a mathematical approach, but the principles can be understood without using advanced mathematics.

Most genetic association studies are population based and use several principles of population genetics, sometimes implicitly, in their design, analysis, and interpretation. This chapter reviews the concepts of population genetics that are relevant to genetic association studies. No mathematical background is assumed, and no such details are provided, but excellent resources are available for more mathematically oriented readers. Familiarity with the concepts discussed in this chapter is essential for anyone planning on conducting genetic association studies.

3.1 Hardy–Weinberg Equilibrium

Genetic variation dynamics can be expressed mathematically

A mutation introduces a new variant to the population, which may or may not be transmitted to the next generation. Even without a new mutation, variant frequencies may still change. It is only natural that if certain people only mate with certain people based on some characters—for example, tall people only with tall people—then any genetic variant that is associated with tallness would be enriched in their offspring. Therefore, nonrandom mating may alter variant frequencies in future generations. In a different scenario, some environmental changes may result in certain variants conferring an advantage or disadvantage to their bearers. If food becomes unavailable, for example, any genetic variants that increase absorption of nutrients above their average absorption rate would confer an advantage for survival and reproduction, so that these variants would be more common in following generations. This is the proposed mechanism for the high frequency of the *HFE* mutation C282Y, which increases iron absorption. Likewise, genetic variants such as *HLA-B*57*, which confers resistance to HIV infection, gain greater frequencies in places where HIV has a high infection rate. Another reason for changes in genetic variation is the introduction of new variants by the addition of new people with a different genetic background to a population. Thus, mutation, selection, nonrandom mating, and migration can generally cause alterations of genetic variant frequencies in populations.

The principle proposed independently by G. H. Hardy and W. Weinberg was based on a very simple approach: if no evolutionary pressure that alters allele frequencies applies to a population, genotype frequencies remain stable from one generation to another.

When this is the case, the population is said to be in **Hardy–Weinberg equilibrium (HWE)** and the proportions of genotypes are directly determined by frequencies of alleles. This relationship between allele frequencies and genotype frequencies can be expressed using a simple mathematical formula that has useful applications. If the major allele (A) of a SNP with two alleles has a frequency p, the minor allele (B) frequency would be $(1 - p)$, because there are two alleles and their total $(p + q)$ should be 1.0. It follows that the square of $(p + q)$ is also 1.0 because the square of 1.0 is 1.0:

$$p + q = 1.0 = (p + q)^2$$

Since p and q represent the allele frequencies of A and B, p^2 in this formula corresponds to the frequency of homozygosity for the common allele A, $2pq$ to the heterozygosity rate (AB), and q^2 is the homozygosity rate for the minor allele B (**Figure 3.1**). This makes intuitive sense because the homozygosity rate is the probability of having allele A on both chromosomes and its frequency is equal to the probability for its presence on each chromosome. The probability that it is present on both chromosomes can be calculated by multiplying the allele frequency by itself. Thus, if allele A has a frequency of p, which denotes the probability of having allele A on one chromosome, then the genotype AA will have a frequency of p^2, which corresponds to having allele A on both chromosomes $(p \times p)$. Just as the two allele frequencies $(p + q)$ sum to 1.0, so the frequencies of their three genotypes AA, AB, and BB, with frequencies p^2, $2pq$, and q^2, respectively, sum to 1.0 as well:

$$p^2 + 2pq + q^2 = 1.0 = p + q$$

The frequency of pq is doubled because genotype AB is equivalent to genotype BA.

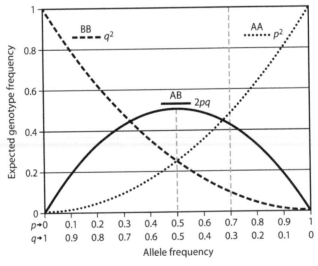

Figure 3.1 **Relationships between allele (A, B) and genotype (AA, AB, BB) frequencies in an equilibrium state.** When allele frequencies for A and B are p and q, the genotype frequencies can be predicted if the population is in Hardy–Weinberg equilibrium. If, for example, allele A frequency (p) is 0.7, allele B frequency (q) is 0.3, and genotype frequencies for AA (p^2), AB ($2pq$), and BB (q^2) will be 0.49, 0.42, and 0.09, respectively. If there are evolutionary forces at work, observed genotype frequencies will show deviations from expected frequencies. Note that the maximum value that $2pq$ (genotype AB frequency) can have is 0.50. Thus, if genotype AB (heterozygosity) frequency for a SNP with two alleles is greater than 0.50, this is evidence of Hardy–Weinberg disequilibrium.

The Hardy–Weinberg principle states that in an infinitely large population where random mating takes place, allele and genotype frequencies remain stable as long as there is no selection, mutation, or migration. The principle can be used to test whether these assumptions are met in a population. The genotype frequencies can be checked against the expected frequencies calculated from the allele frequencies by a simple statistical test called the goodness-of-fit test. If the expected and observed genotype frequencies are identical or similar, a statistical assessment will show a good agreement (fit) between the observed and expected frequencies, and the population is assumed to be in HWE. If there are statistically significant differences between the observed and expected frequencies, then the population is assumed to be in Hardy–Weinberg disequilibrium (HWD). HWD is observed if one of the assumptions of the equilibrium (mutation, random mating, selection, migration) is violated. In practice, however, the most common reason for violation of equilibrium is not biological. If a healthy population sample shows signs of HWD, errors in obtaining genotypes is the most plausible explanation. For this reason, HWE testing has established itself as a genotyping quality control test in the context of a genetic association study.

Typically, HWE is tested in unrelated individuals using a goodness-of-fit test that compares expected and observed numbers of genotypes. While observed frequencies are obtained by genotyping individuals, expected frequencies are calculated by using observed frequencies. A worked example of the calculation of expected frequencies is given in **Box 3.1**. First, allele frequencies are calculated from the three genotype frequencies. These allele frequencies are then converted to expected genotype frequencies using the Hardy–Weinberg principle. It is then a matter of checking the agreement between observed and expected genotype frequencies. Note that the allele frequencies used in the estimation of expected frequencies are derived from observed genotype frequencies. This may sound counterintuitive, but the idea is to check whether these alleles have formed the genotypes as randomly as they should.

The goodness-of-fit test used to test HWE is not a powerful test and the reliability of the result rests on the sample size. The genotype proportions in the example given in Box 3.1 part B are used to illustrate the importance of sample size, as also shown in **Table 3.1**. If HWE is found to be violated, regardless of the sample size, it is a valid result, but HWE may still be violated even if the statistical result does not suggest so. This interpretation is important when using HWE for genotyping quality control, as will be discussed in Chapter 7.

There is a rule of thumb that can be applied as a test of HWE in any sample, whatever the size. Figure 3.1 shows genotype frequencies (the curves) for the whole range of allele frequencies (p and q in the x axis). It shows that the AB (heterozygote) genotype frequency can only have a maximum value of 0.50, and it does so when the individual allele frequencies are both 0.5 ($p = q = 0.5$, therefore $2pq = 0.50$). It follows that if the heterozygote frequency is found to be more than 0.50, something is wrong. An HWE violation is also indicated if the heterozygosity rate is 0.50 (or close) but the allele frequencies are not close to 0.5. The genotype frequencies given in Table 3.1 are 0.42 (AA), 0.50 (AB), and 0.08 (BB), but the heterozygote (AB) frequency should only be 0.50 when the allele frequencies are also 0.50, and the genotype frequencies AA and BB are therefore 0.25. However, the observed frequencies are 0.42 and 0.08. This discrepancy between expected and observed frequencies only shows as a lack of goodness-of-fit result when the sample size is large enough.

Box 3.1 Worked examples of calculating expected frequencies

(A) Observed genotype counts (proportions):

AA = 36 (0.360)

AB = 48 (0.480)

BB = 16 (0.160)

Sample size (n) = 100

Allele A count = 2 × (AA count) + 1 × (AB count) = 2 × 36 + 48 = 120

Allele B count = 2 × (BB count) + 1 × (AB count) = 2 × 16 + 48 = 80

Total chromosome number ($2n$) = 200

Allele A frequency (p) = 120/200 = 0.600

Allele B frequency (q) = 80/200 = 0.400

Expected genotype proportions:

p^2 = 0.6 × 0.6 = 0.360

$2pq$ = 2 × 0.6 × 0.4 = 0.480

q^2 = 0.4 × 0.4 = 0.160

These frequencies are identical to the observed frequencies. Therefore, the fit between the observed and expected frequencies is perfect and the sample seems to be in HWE.

(B) Observed genotype counts (proportions):

AA = 168 (0.420)

AB = 200 (0.500)

BB = 32 (0.080)

Sample size (n) = 400

Allele A count = 2 × (AA count) + 1 × (AB count) = 2 × 168 + 200 = 536

Allele B count = 2 × (BB count) + 1 × (AB count) = 2 × 32 + 200 = 264

Total chromosome number ($2n$) = 800

Allele A frequency (p) = 536/800 = 0.670

Allele B frequency (q) = 264/800 = 0.330

Expected genotype proportions:

p^2 = 0.670 × 0.670 = 0.449

$2pq$ = 2 × 0.670 × 0.330 = 0.442

q^2 = 0.330 × 0.330 = 0.109

These frequencies are different from the observed frequencies. Therefore, the fit between the observed and expected frequencies is not good. The statistical assessment by a goodness-of-fit test reveals a statistically significant difference between observed and expected frequencies. The sample seems to be in HWD.

Table 3.1 HWE goodness-of-fit test results for the same
genotype proportions in different sample sizes

Sample size	50	100	200	300	400
AA count	21	42	84	126	168
AB count	25	50	100	150	200
BB count	4	8	16	24	32
HWE test result (P)	0.52	0.26	0.08	0.02	0.009

The genotype proportions are the same throughout—AA is 0.420, AB is 0.5, and BB is 0.080—
but the goodness-of-fit results differ with sample size. The test result must be less than 0.05 to be
statistically significant. As the sample size increases, the test results begin to indicate a significant
difference. This phenomenon is due to the statistical power of a test and will be discussed in detail
in Chapter 6.

3.2 Linkage Disequilibrium

Disequilibrium begins where equilibrium ends

Linkage disequilibrium (LD) is concerned with the independence of allelic combinations in haplotypes. If certain combinations occur at a different frequency from the expected, then they are not independent of each other and have some relationship which brings them together more or less frequently than expected. The state of total independence—a random combination of alleles at different SNPs—is called equilibrium. Any deviation from the equilibrium state is therefore disequilibrium. In equilibrium, a haplotype's frequency can be estimated from the allele frequencies: since they are independent, the multiplication product of the two allele frequencies gives the haplotype frequency. In disequilibrium, there is a lack of independence, and the haplotype frequency is different from the product of the allele frequencies.

To understand the relationships between SNPs, it is best to begin with how SNPs and their haplotypes are formed (**Figure 3.2**). When a SNP is present in the chromosome, its alleles form two haplotypes with a nearby nucleotide that is constant. When one of the constant nucleotides mutates, this gives rise to a third haplotype. In following generations, exchange of chromosomal segments due to recombination events generates a fourth haplotype. If there is no advantage for an allelic combination in a haplotype, the alleles remain independent and the haplotype frequency is easy to estimate. If the observed haplotype frequency is different from the estimated (expected) value, it can be concluded that there is some reason (usually a selective advantage or disadvantage) for the two alleles to be on the same chromosome more or less often than expected.

As an example (**Figure 3.3**), if one allele of SNP1 has a frequency of 0.800 and one allele of SNP2 has a frequency of 0.200, then the probability that these two alleles will be on the same chromosome is $0.800 \times 0.200 = 0.160$. This is the expected frequency for their co-occurrence on the same chromosome (haplotype). In an equilibrium state, the two alleles are randomly transmitted from parents to offspring and they behave within the limits of statistical independence, so that the presence of one allele does not depend on the presence or absence of the other allele on the chromosome. If two alleles from different loci tend to occur on the same chromosome more frequently than expected (for example, 0.240 rather than 0.160 in this example; see Figure 3.3), then they are in positive linkage disequilibrium. If they occur less frequently than expected (observed frequency < expected

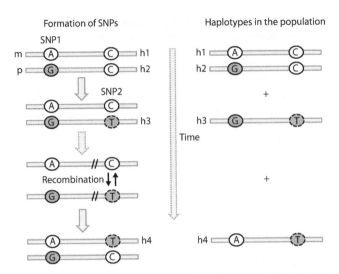

Figure 3.2 Formation of a new SNP on a chromosome leading to new haplotypes in the population. SNP1 already exists in the population, resulting in haplotype 1 (h1) and haplotype 2 (h2). When SNP2 is formed due to a mutation, a new (GT) haplotype is formed (h3), which can eventually give rise to a fourth haplotype (h4) due to recombination. Note that the new T allele of SNP2 will always be associated with allele G of SNP1 until a recombination event breaks this exclusive relationship. Until then, alleles G and T will always be on the same haplotype in this population. These two alleles are said to be linked, and this non-independence will generate linkage disequilibrium.

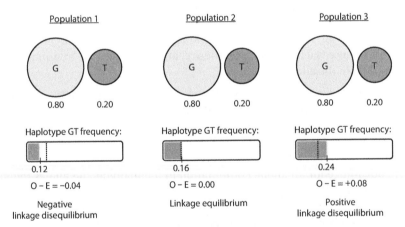

Figure 3.3 Linkage disequilibrium can be negative or positive. In three different populations, the frequencies of allele G at SNP1 and allele T at SNP2 are 0.80 and 0.20, respectively. This gives an expected (E) frequency of 0.16 (0.80 × 0.20) for the haplotype GT. Observed (O) or actual frequencies in the three populations are 0.12, 0.16, and 0.24, respectively. The difference between observed and expected frequencies (O − E) is −0.04 in population 1, 0.00 in population 2, and +0.08 in population 3. Therefore, allele G of SNP1 and allele T of SNP2 are in linkage equilibrium in population 2 (their observed haplotype frequency is equal to the expected frequency, indicating independence); in population 1, they are in negative linkage disequilibrium (the haplotype frequency is less than the expected value, indicating a tendency not to be on the same chromosome, presumably due to some disadvantage that they confer when together in the same chromosome); and in population 3, they are in positive linkage disequilibrium (the haplotype frequency is more than the expected value, indicating a tendency to be on the same chromosome, presumably due to some advantage that they confer when together in the same chromosome).

frequency), they are in negative linkage disequilibrium. The differences between the observed and expected frequencies form the basis of the LD measures used to quantify the magnitude of disequilibrium.

Linkage disequilibrium decays through repeated recombination events

When a nucleotide substitution occurs on a chromosome for the first time, there are already other polymorphisms around it. The new nucleotide is in complete linkage with other alleles on the same chromosome (see haplotype h3 in Figure 3.2). This initial state immediately after a new mutation occurs on the chromosome is the disequilibrium state. If this state is maintained, it is possible to know which alleles of nearby SNPs are present on the chromosome given the allele present in one of them. LD is maintained until neighboring polymorphisms are separated by **recombination** or crossing-over events in following generations (see haplotype h4 in Figure 3.2). Until a recombination event, the new nucleotide in the new polymorphism will always be on the same chromosome as the nucleotides that were present when it first occurred. Recombination is a natural phenomenon that occurs during meiosis and results in the exchange of small segments between homologous chromosomes (**Figure 3.4**). Meiosis occurs once per generation (in the formation of gametes), so a recombination event is possible once per generation. Linkage disequilibrium therefore decays as a function of the number of generations since the occurrence of a new mutation.

The first recombination event may not occur for several generations, by which time the chromosome with the new variant may have already spread in the population. After recombination has occurred, a mixture of chromosomes will be present in the

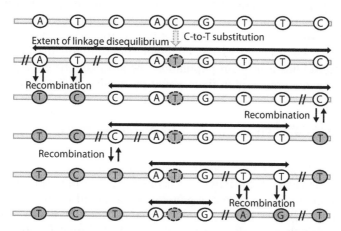

Figure 3.4 Linkage disequilibrium decays over generations due to repeated recombinations. In this example, the initial mutation is a C-to-T substitution. When a mutation occurs on a chromosome, the new allele is in linkage disequilibrium (LD) with all other alleles on the same chromosome. The extent of LD is indicated by the horizontal arrow above the chromosome. In successive generations, exchange of chromosomal regions (recombinations) during meiosis on either side of the mutation replaces the alleles at neighboring SNPs, and the extent of LD between the mutant allele T and the alleles of the original chromosome starts to diminish. As time passes, the extent of LD continues to diminish and only the nearest markers remain in LD with the new variant, and continue to be linked during transmission from one generation to the next. Ultimately, even the nearest alleles may be separated from the mutant allele T; then, not being linked to any other allele in the chromosome, allele T will reach an equilibrium state.

population: the haplotypes that were present before the mutation occurred (h1 and h2 in Figure 3.2); the haplotype consisting of the mutant allele flanked by the allele(s) originally on the same haplotype (h3 in Figure 3.2); and a new haplotype consisting of the mutant allele flanked by different alleles due to the transfer of the mutant allele via recombination (h4 in Figure 3.2). The ability to absolutely predict the alleles in flanking SNPs will be reduced. Since this diversity may further increase in each subsequent generation through further recombinations (see Figure 3.4), predicting the alleles of flanking SNPs becomes more and more difficult over time. Ultimately, repeated recombinations will break any relationship between neighboring SNPs and they will become totally independent from each other: they will be in equilibrium. Once this state is reached, it is not possible to predict the alleles of flanking SNPs. The probability of having a recombination event is obviously greater if SNP pairs are separated by a longer distance. Thus, SNP pairs that have longer distances between them tend to be closer to the equilibrium state and have lower LD than SNPs closer to each other.

Average LD is lower in populations that have been stable for long periods of time, mainly because there have been more meiosis events and recombinational exchanges within that population. LD is therefore lower in African populations compared with other populations. Any ancestral small African subpopulation that came out of Africa consisted of relatively few people with smaller diversity than that of the whole African population at that time. Each of the ancestral population movements out of Africa can be seen as resetting the clock for the occurrence of LD in the new population. Existing LD patterns will be maintained until they are broken in successive generations. It follows that populations that have just been through a contraction and expansion phase, known as a **bottleneck**, also have a high level of LD. Likewise, LD should be lower in a fast-growing population, compared with populations growing more slowly, due to increased opportunities for recombinational events (reproductive cycles) in the same time interval. All of these properties of populations that influence LD levels have implications in genetic association study design, as will be discussed in later chapters. For example, since LD is lower in African populations, studies in African populations need to genotype a higher number of SNPs for the same genomic region than would be needed in the study of a European population.

Linkage disequilibrium can also occur through mixing with a new population

Another way LD occurs is through the introduction of new variants to a population by recent addition of a new subpopulation. When a new admixed population is formed, a high level of LD between unlinked markers is observed. It is important to recognize this type of population admixture, as it can generate widespread LD in the genome, which complicates interpretation of genetic association study results. It is a mathematical complication rather than these markers really being in LD. Fortunately, obvious population admixture can now be recognized using molecular markers, and such complications can be largely avoided.

Why is linkage disequilibrium important?

Genetic association studies aim to identify the primary causal SNP modifying disease susceptibility or affecting a quantitative trait (**direct association**) but, most of the time, the associations found are with nearby polymorphisms that are in LD with the unknown causal SNP (**indirect association**). In studies mapping a disease susceptibility locus, LD graphs may be produced that show the change of LD values between pairs of SNPs across a candidate region in order to narrow down a subregion for detailed analysis. The causal locus is most likely where these pairwise LD values for adjacent markers reach a peak in diseased individuals.

By virtue of most of them being in strong LD with the causal SNP, most SNPs in the region flanking the causal SNP are in strong LD with one another (**Figure 3.5**). The area showing high LD is then screened for more polymorphisms and sequencing studies are performed to identify the causal SNP. This is called LD mapping. In practice, therefore, we rely on LD between known markers and the unknown causal variant to locate the disease locus, and then zoom in to find out the exact disease-causing variant. The same principle also guided the classic linkage studies (which are not covered by this book). Linkage studies examine known markers in families with affected members to find out the location of a marker that is in linkage within a family with the unknown causal disease locus.

In the SNP selection phase of a study, it is important to avoid redundancy so as not to waste resources. If all variants in a gene are included in a study, there is a chance that most of them will be highly correlated with each other due to LD, and will not provide additional information for analysis over the one SNP that represents all others correlated to it. Also, if SNPs that are correlated with each other are included, all of them will show associations, but these will not be independent associations. It is therefore crucial to run an LD analysis in SNP selection in order to choose independent markers. There are many algorithms developed to select the most informative subset of a group of SNPs. Some of these mathematical approaches will be introduced in Chapter 8.

There are several measures of linkage disequilibrium

Since LD correlates with the probability of two alleles being on the same chromosome, the simplest measure of LD is the difference between the observed and expected values of haplotype frequencies. This is called the delta (Δ) value (D for difference in the Greek alphabet). The observed value is estimated from either family- or population-based data, and the expected value is calculated from allele frequencies estimated from observed data. If the frequency of a haplotype formed by variant alleles of two adjacent SNPs is

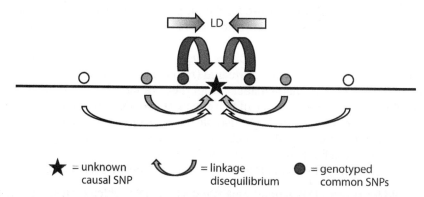

★ = unknown causal SNP   ⌣ = linkage disequilibrium   ● = genotyped common SNPs

Figure 3.5 Linkage disequilibrium patterns around the causal variant. Since LD decays with distance, the closer a SNP is to the causal SNP, the stronger the LD (curved arrows) between them. The SNPs in strong LD with the unknown causal LD will also be in strong LD with one another (as shown by the straight arrows). This LD among genotyped SNPs will get stronger nearer the location of the unknown causal SNP in affected individuals, but not in healthy controls who do not possess the causal SNP.

0.125, and the frequencies of the two alleles are 0.250 and 0.300, the Δ value is calculated as follows:

$$\Delta = \text{observed frequency} - \text{expected frequency}$$

$$= 0.125 - (0.250 \times 0.300)$$

$$= 0.125 - 0.075$$

$$= 0.050$$

This result suggests that the two alleles tend to occur more frequently than expected, and indicates positive LD; that is, these two alleles have not yet been separated by repeated recombinations over many generations. It follows that if the observed frequency is less than the expected frequency, there would be negative LD, which has a different interpretation. Negative LD means that the two alleles tend to be on the same chromosome less frequently than expected. It may be that their co-occurrence has some deleterious effects and their combination has been selected against. Alternatively, the negative LD may just be the mirror image of positive LD by alternative alleles at those two polymorphisms.

The simplest measure of LD (Δ) makes intuitive sense, but it is not widely used because of its sensitivity to allele frequencies: the same magnitude of difference between observed and expected haplotype frequencies will yield a much larger value for common alleles compared with rare alleles. The example given above yielded a Δ value of 0.050, but, if the allele frequencies are one-tenth of 0.250 and 0.300 (that is, 0.025 and 0.030), the Δ value would be 0.005. A simple mathematical adjustment can be performed to normalize the Δ value, giving it the same magnitude for different allele frequencies. The normalized Δ value is called D'.

As LD decreases with the number of recombinations since the emergence of a new variant, it follows that if all four possible chromosomes are observed, LD will be weak. The relative magnitude of D' is not easy to interpret, and D' is most useful when its value is 1.0, denoting no recombination between the two alleles since the more recent one appeared. No recombination results in the total absence of one of the four possible haplotypes.

The most commonly used LD measure for the purposes of genetic association studies is neither Δ nor D'. Since strong LD allows prediction of one allele in a haplotype from the other allele, and this means a strong correlation between the two, a correlation coefficient should provide valuable information. For the purposes of genetic association studies, the square of the correlation coefficient, r-squared (r^2), is the most commonly used LD measure and has very favorable mathematical properties (**Table 3.2**). This value quantitates the correlation between the presence of two alleles at two SNPs. In the perfect LD situation ($r^2 = 1$), one allele is completely predictable once the other one is known. This is why an alternative term for LD is allelic association. LD is also referred to as gametic association, since gametes are haploid and alleles with absolute LD will always be in the same gamete together. Some features of LD in the human genome are listed in **Box 3.2**.

The threshold for strong LD is generally accepted as $r^2 \geq 0.80$, which corresponds to a correlation coefficient of approximately 0.90. As the intermediate values of r^2 are proportional to the degree of correlation, it measures how well one marker can act as a surrogate, or proxy, for another. The r^2 value can be used to estimate the sample size in a genetic association study. Most associations found are with a SNP nearby and are in LD with an unknown causal locus. Suppose the r^2 value between a causal disease locus and a nearby marker

Table 3.2 Properties of commonly used bi-allelic LD measures

LD measure	Features
Δ (delta) (no range)	Used in earlier studies when more sophisticated LD measures were not available or could not be computed Strongly influenced by allele frequencies Cannot be used to compare LD quantities among different pairs of alleles when allele frequencies are different
D′ (D-prime) (normalized Δ; range 0 to 1)	Useful in inference of recombinational history. D′ = 1 when there has been no recombination between the two alleles (and only three of four possible gametes are observed) D′ < 1 is evidence for historical recombination Values between 0 and 1 are hard to interpret Small sample size and rare alleles inflate D′ D′ may be 1 despite a low r^2 Limited value in comparisons
r^2 (r-squared) (range 0 to 1)	When the two alleles are always on the same chromosome, and are exclusively associated, $r^2 = 1$ Intermediate values are easy to interpret Small sample size does not inflate the value Robust for allele frequency differences Most valuable for comparison between studies, for tagSNP selection, and power calculations

locus is known. The lower the r^2 value, the greater the sample size will have to be to detect the association with the surrogate of the causal locus. If $r^2 = 1$, all chromosomes that carry the causal allele will also carry the surrogate marker, but as the correlation (LD) between them decreases, fewer people will carry the surrogate marker on the same chromosome as the causal disease locus. A larger sample size will therefore be needed to detect the association of a surrogate marker. Assume that a study shows an association with the causal disease allele with a given statistical significance. To find an association at the same statistical significance level, but using a surrogate marker, the sample size has to be increased by a factor of $1/r^2$. Of course, the causal locus is not generally known, and a range of power calculations are made for different r^2 values, reflecting the strength of the candidate polymorphisms, to select a sample size. One can be confident that average r^2 values for LD between markers used in a genetic study and the unknown causal allele increase as more and more markers are included in the study because of the concept illustrated in Figure 3.5.

How the two most popular measures of LD, D′ and r^2, correspond to each other is not easy to discern. Depending on the allele frequencies at both loci, and whether all four possible haplotypes are present in the population, there may or may not be any correlation between them. It is more common to have a D′ value of 1.0 but a low r^2 value than the opposite way around.

Box 3.2 Linkage disequilibrium in numbers

Linkage studies identify linkage areas of 10–20 Mb that contain the causal variant.

Haplotype blocks characterized by high LD among its variants in the human genome are on average ~20 kb in size.

The average range over which variants show LD is between 60 and 200 kb in general populations.

The most extensive LD has been reported for the most common European HLA haplotype, HLA-A1B8DR3, which shows LD extending to distances beyond 1 megabase.

Table 3.3 Table to test the significance of LD results

	SNP2 allele A	SNP2 allele C
SNP1 allele A	530	86
SNP1 allele C	384	0

Regardless of which LD measure is used, statistical significance testing for LD is needed. This is performed using a table (**Table 3.3**) that records the co-presence of the alleles at the two SNPs (corresponding to possible gametes). If the co-presence of alleles is random, the difference between the rows and columns will not be statistically different. If there is LD (the alleles are not independent), the difference will be statistically significant. The example shown in Table 3.3 is based on a sample size of 1000 chromosomes and there seems to be an imbalance in the pairing of alleles (increased co-occurrence of alleles A and A, but against co-occurrence of alleles C and C), suggesting LD between alleles A (SNP1) and A (SNP2).

Here, the discussion of LD and its measures is restricted to bi-allelic LD only, but it is possible to calculate LD for multiple markers. Such multi-locus LD measures require sophisticated algorithms and are beyond the scope of this book.

3.3 Population Substructure

Populations are usually heterogeneous in terms of genetic ancestry, with people of different ancestral origins living together in that population. Sometimes this is obvious, and sometimes it may be more cryptic. When it is clear that people from different strata, such as ethnicities, are included in a sample, the situation is called either **population substructure** or **population stratification**. When it is more subtle, the situation is known as **cryptic population structure**. Either way, this phenomenon may confound genetic association study results, leading to false-positive or false-negative results. The problem arises when the disease is more common in one subpopulation and marker frequencies are also different among subpopulations. SNPs with no allele frequency differences between subpopulations will not show any spurious associations due to population stratification. SNPs that show a spurious association in a study that suffers from the population substructure problem will have no link with the disease, but will show the association due to a mathematical complication.

Admixed populations allow admixture mapping for risk marker identification

An admixed population is one in which distinct ancestral origins have been recently mixed. A typical example is Hispanics in the Americas, who harbor various proportions of European, West African, and Indigenous American ancestry in individual genomes. On average, Hispanics in the Dominican Republic have a greater proportion of African contributions to their genome, while Hispanics in Mexico have more Indigenous American contributions. If a disease shows different incidence rates in two populations, and if these populations have contributed to an admixed population, a statistical methodology called **admixture mapping** is used to detect genetic variants that are responsible for ethnic differences in disease risk. Detection is possible because there will be continuing LD between ancestral markers unique to the ancestral population and a causal disease locus for a disease that has a higher incidence rate in one ancestral population.

Genetic admixture can be exploited when there are markers for each ancestry. Such markers for each major population group are called **ancestry informative markers** or **AIMs**.

By using these markers and specially developed algorithms, the genomic proportions of each ancestry in an individual can be calculated. These proportions can be treated as a continuous variable and correlations sought with disease susceptibility. One such study has identified a strong correlation between the proportion of European ancestry and breast cancer risk among US Hispanics. A follow-up study used the admixture mapping approach to identify which genomic regions that originated from the same genetic ancestry were shared by the cases but not by healthy controls. Basically, the first step seeks associations between AIMs and disease, and the second step elaborates this association by examining candidate polymorphisms near that signal to identify the disease locus.

The advantage of admixture mapping originates from the stronger LD extending over greater distances in recently admixed populations. Large stretches of chromosomes unique to the ancestral population that have the disease locus have not had sufficient time to recombine with chromosomes from the other ancestral population without the disease marker. For diseases that vary in prevalence between two or more ancestral populations, this long-range LD is exploited to search for genetic markers of the ethnic difference in disease risk (**Figure 3.6**). The strong LD (correlation) that is exploited is between AIMs and the unknown causal disease locus. In admixed populations, AIMs for the high-risk population in LD with a locus responsible for an ethnic difference in disease risk will have a greater-than-expected proportion of disease cases. The genomes of diseased individuals

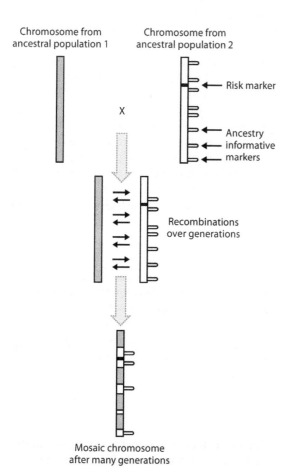

Figure 3.6 The basis of admixture mapping. When two ancestral populations are mixed, over subsequent generations, their chromosomes form hybrid chromosomes or mosaics due to the exchange of segments via recombinations. If the ancestral population 2 has a higher frequency of the disease and the marker is flanked by ancestry informative markers (AIMs), the risk marker is tagged by AIMs. By working out AIM allele frequencies across the genome in people with the disease, the segments that have come from the ancestral population 2 can be identified. Note that the segment with the disease marker will be enriched in people with the disease and can be identified easily with the increased frequency of AIM alleles around that segment.

are screened for the proportion of markers indicating the ancestry that has the highest disease risk, and regions with high proportions of such markers and that may harbor a susceptibility gene are identified. This approach localizes the unknown disease locus to a more manageable, narrow genomic region, just as the classic linkage studies do, but without the need for pedigrees. Since this approach relies on LD in admixed populations, it is also called **mapping by admixture linkage disequilibrium** (**MALD**). Compared with other types of association studies, MALD requires fewer markers for mapping susceptibility loci.

3.4 Signatures of Natural Selection

The aim of genetic association studies is to discover disease-causing variants in order to learn about disease pathogenesis and aid development of preventive and therapeutic interventions. Disease-causing variants cause changes in physiology and are therefore subject to selective pressures. It follows that a genomic region that contains signals for natural selection may contain a causal disease locus. Natural selection increases the transmission of advantageous alleles (positive selection) and operates against the transmission of deleterious ones (negative or purifying selection). The classic example given is a very visual one and concerns the increased frequency of black moths during the industrial revolution in England. Increasing environmental pollution during the industrial revolution provided an advantage for black moths in favor of white ones, in terms of better camouflage and protection from predators. As a result, black moths increased in frequency during that time. Likewise, it has been suggested that the *CCR5* mutation Δ32 has increased in Northern Europe because of its protective effect against infectious diseases such as bubonic plague, smallpox, and West Nile virus. The mutation that causes sickle-cell anemia has high frequency in malarial regions because it confers protection from malaria. The type of hemoglobin caused by the mutation has decreased capacity to carry oxygen and it increases the tendency of red cells to assume a different shape, but these changes create an unfavorable environment for the malaria parasite to survive. Since malaria is a more common disease than sickle-cell anemia, the protection from malaria has a stronger selective effect and the mutation survives in those regions where malaria otherwise kills a lot of people. Genetic variants that have been subject to selective pressure show signatures in the genome that can be recognized by certain algorithms. If a choice between two markers has to be made for inclusion in a study, choosing one that appears to be in a region that shows signs of selection is preferable.

The most commonly used statistical tests to examine signatures of selection in population-level genetic data are: F_{ST} (a measure of the degree of genetic differentiation of subpopulations); the integrated haplotype score (iHS); the GERP (genomic evolutionary rate profiling) score; the RS (rejected substitution) score; and extended haplotype homozygosity (EHH). These tests have yielded information on the correlation between heritable disease susceptibility and natural selection; this information has been embedded into bioinformatics databases and each variant can be checked for its selective value. The details will not be provided here, but the published literature is readily available for detailed studies and information.

Key Points

- Hardy–Weinberg equilibrium indicates that no evolutionary forces act on a population. In such populations, genotype frequencies can be estimated directly from allele frequencies. Any deviation of observed genotype frequencies from expected frequencies indicates the action of evolutionary forces.

- When the genotype frequencies of a SNP show disagreement with expected frequencies, it may be an indication of selection acting on this SNP. Alternatively, such a finding may point to a genotyping error.

- One simple rule of thumb that derives from the Hardy–Weinberg principle is that the heterozygosity rate for two SNP alleles cannot exceed 0.500, whatever the allele frequencies. Any heterozygosity rate above 0.500 should be examined carefully.

- The concept of linkage disequilibrium is the foundation of disease-gene mapping. A rare and unknown causal disease locus can be mapped due to its LD with common variants nearby that show associations with disease.

- Linkage disequilibrium forms the basis of SNP selection by helping to leave out SNPs that are correlated with one another and do not provide any additional information for genetic association studies.

- Population substructure is not always obvious and when it exists in a cryptic form, it creates complications for genetic association studies.

- Signatures of selection in the human genome are recognizable and such regions are likely to harbor functional, disease-causing variants.

URL List

EvoTutor Online Simulations. http://www.evotutor.org

Holsinger K. Population Biology Simulations. http://darwin.eeb.uconn.edu/simulations/simulations.html

Further Reading

Population genetics

Jobling M, Hollox E, Hurles M et al. (2013) Human Evolutionary Genetics, 2nd ed. Garland Science. (*In particular, Chapter 5 "Processes Shaping Diversity" provides further details of topics covered in this chapter including HWE and the evolutionary forces that disturb HWE, and selection.*)

Hardy–Weinberg equilibrium

Salanti G, Amountza G, Ntzani EE & Ioannidis JP (2005) Hardy-Weinberg equilibrium in genetic association studies: an empirical evaluation of reporting, deviations, and power. *Eur J Hum Genet* 13, 840–848 (doi: 10.1038/sj.ejhg.5201410). (*Comprehensive discussion of HWE in the context of genetic association studies and especially the statistical power issue.*)

Wang J & Shete S (2012) Testing departure from Hardy-Weinberg proportions. *Methods Mol Biol* 850, 77–102 (doi: 10.1007/978-1-61779-555-8_6). (*Provides background and a step-by-step guide on testing HWE.*)

Linkage disequilibrium

Mueller JC (2004) Linkage disequilibrium for different scales and applications. *Brief Bioinform* 5, 355–364 (doi: 10.1093/bib/5.4.355). (*An excellent review of LD and its measurement as well as multi-locus LD.*)

Reich DE, Cargill M, Bolk S et al. (2001) Linkage disequilibrium in the human genome. *Nature* 411, 199–204 (doi: 10.1038/35075590). (*A survey of LD in the human genome in different populations that makes links between LD and population history.*)

Population substructure

Cardon LR & Palmer LJ (2003) Population stratification and spurious allelic association. *Lancet* 361, 598–604 (doi: 10.1016/S0140-6736(03)12520-2). (*Reviews the concept of population structure, and problems associated with it and their remedies.*)

Admixed populations and admixture mapping

Fejerman L, Chen GK, Eng C et al. (2012) Admixture mapping identifies a locus on 6q25 associated with breast cancer risk in US Latinas. *Hum Mol Genet* 21, 1907–1917 (doi: 10.1093/hmg/ddr617). (*A good example of using admixture mapping to identify a risk marker for a disease that shows correlation with different ancestral groups in its incidence rates.*)

Admixture mapping

Montana G & Hoggart C (2007) Statistical software for gene mapping by admixture linkage disequilibrium. *Brief Bioinform* 8, 393–395 (doi: 10.1093/bib/bbm035). (*Although it covers statistical aspects, this paper concisely reviews admixture mapping for non-statisticians.*)

Natural selection

Crisci JL, Poh YP, Bean A et al. (2012) Recent progress in polymorphism-based population genetic inference. *J Hered* 103, 287–296 (doi: 10.1093/jhered/esr128). (*An advanced-level discussion of polymorphism-based approaches in population genetics covering detection of selection and population demographics. Mathematical and statistical approaches are presented in sufficient detail.*)

Hughes AL, Welch R, Puri V et al. (2008) Genome-wide SNP typing reveals signatures of population history. *Genomics* 92, 1–8 (doi: 10.1016/j.ygeno.2008.03.005). (*Exemplifies how genome-wide polymorphism data can be used to extract information in population genetics with examples of how this information helps with genetic association study design and interpretation.*)

Oleksyk TK, Smith MW & O'Brien SJ (2010) Genome-wide scans for footprints of natural selection. *Philos Trans R Soc Lond B Biol Sci* 365, 185–205 (doi: 10.1098/rstb.2009.0219). (*Provides theoretical background on natural selection, statistical tests, and actual examples of selection identified by genome scans.*)

Epidemiologic Principles and Genetic Association Studies

4

This chapter will discuss the epidemiologic and statistical concepts that are of importance in conducting a successful genetic association study. These concepts are used in the design, execution, analysis, and interpretation phases. The information contained in this chapter is not meant to be a substitute for having an epidemiologist or biostatistician in the research team, but to allow fruitful interactions among the members of research teams.

4.1 Types of Epidemiologic Association Studies

Epidemiologic studies involving an interventional arm are known as **interventional** studies, and are represented by randomized clinical trials. Noninterventional studies are known as **observational** studies and mainly consist of **cohort** studies or case-control studies. Another observational study type are **cross-sectional** studies, which provide a snapshot at any given time point. Cohort studies begin with a large, healthy population sample and both **exposure** and **outcome** are recorded as they occur during **longitudinal** follow-up. This characteristic of cohort studies makes them **prospective** studies. Case-control studies, however, begin with cases who have already developed the phenotype of interest (outcome) and collect the exposure information retrospectively. Thus, case-control studies are **retrospective** studies. When subjects are recruited prospectively and independently of their disease status, a long-term follow-up allows direct estimation of the risk. In retrospective studies like case-control studies, however, the disease risk cannot be estimated directly. Instead, some approximation of the risk is calculated.

An exposure may be smoking, a dietary factor, living in a sunny climate, taking a multivitamin supplement, or having irregular sleeping patterns, for example. An outcome is usually a disease condition, but can be any other trait (for example, blood glucose level, body mass index, birth weight, progression from HIV infection to acquired immune deficiency syndrome [AIDS], or surviving a condition). In the most common study design, case-control studies, exposures are surveyed in people with the disease (cases) and without the disease (controls), rather than exposing one group to the potential risk factor and not exposing a control group, as occurs in interventional studies. Such an interventional study may well be unethical in humans and usually can only be done in animals. Common study designs in epidemiologic research are shown in **Figure 4.1**.

Case-control studies have advantages and disadvantages

In retrospective studies, such as case-control studies, cases with the disease have already been diagnosed, and therefore these studies do not require a lot of time to recruit a sufficient number of samples to do the study. This is a very useful feature, especially for rare diseases. However, case-control studies only measure an association, which is simply a correlation, and not a causation. The results of a case-control study do not establish causality, but certain types of prospective studies can. Case-control studies are the most

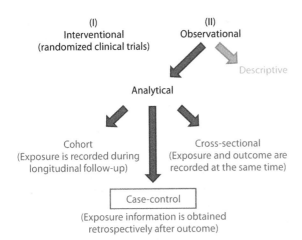

Figure 4.1 Common epidemiologic study designs. Interventional studies are not practical for genetic association testing. The most common design, case-control studies, is an observational study design.

commonly used epidemiologic study design to test a hypothesis. In contrast, a cross-sectional study can only generate a hypothesis.

The strength of an association found in a retrospective case-control study is quantitated by an **odds ratio (OR)**, which approximates to the change in risk of developing an uncommon disease following exposure to a certain variable. For example, the risk of lung cancer is increased tenfold in smokers. The calculation of the odds ratio uses a 2×2 **table** with the cases and controls as rows, and positive or negative exposure as columns (see Chapter 9 for a more detailed explanation). If the odds ratio is less than 1, it denotes a protective association, meaning that the exposure correlates with a decreased risk. An odds ratio greater than 1 denotes increased risk. For common diseases, risk change is estimated by **relative risk** in a cohort study. As ORs are approximations, they should be considered within a range identified by a **confidence interval (CI)**.

A case-control study investigates associations in a disease that is uncommon. The two comparison groups are formed based on the outcome rather than the presence of exposure history, and exposure history is obtained later. It is this fundamental difference from prospective study design that creates a number of drawbacks (**Table 4.1**). It is said that a

Table 4.1 Advantages and disadvantages of case-control studies

Advantages	Disadvantages
Allows execution of easy, quick, and affordable studies	Subject to bias (selection, information [recall], misclassification bias)
	Matching cases and controls is not straightforward
Multiple exposures can be examined for their relationship with the outcome	Direct incidence estimation is not possible; the risk can be approximately calculated as an odds ratio
Rare diseases and diseases with long latency can be studied without waiting for a long time	Multiple outcomes cannot be studied
	If the incidence of exposure is high, it is difficult to show the difference between cases and controls
Suitable when randomization is unethical and when exposure is naturally occurring	Reverse causation is a problem in interpretation (but not in genetic association studies of disease risk)
	Temporal relationship is not clear (except in genetic association studies)
	Causality cannot be established

case-control study begins from the end, as opposed to a prospective study (typically, a cohort study) that begins before the outcome has developed.

One of the most crucial aspects of case-control studies is the selection of controls, which can be a major drawback if not done properly.

Poor selection of controls can lead to spurious associations

Controls should be a randomly selected group from the general population in order to avoid **spurious associations** that cannot be circumvented by statistical manipulations at the analysis phase, and also to avoid results that will not be replicated in subsequent studies. The selection process may be logistically difficult and costly, making reliance on already existing **convenience samples** as controls a common cost-cutting method. As long as the representativeness of a convenience sample is assured, this practice is acceptable. A common convenience sample consists of blood donors or employees/students at the research site, but both of these groups of people may be different from the general population from which the cases come without any selection. For example, most blood donors are regular donors who tend to be taking care of themselves much better than other individuals and undergo tests before giving blood. Similarly, employees are also a healthy subset of the population, and using employees as controls is a well-known source of bias called the **healthy worker effect**. These people may have different exposure profiles compared with the general population. It is therefore better to use a randomly selected healthy control group from the general population, and the drawbacks of any other approach should be recognized.

Healthy controls may include future cases

In a case-control study, controls are selected on the basis of them not having the disease of interest. It may, however, be that some of them will still develop the disease sometime in the future. In this case, they may also possess the **risk markers** for the disease. This situation lowers the ability to detect an association, and causes a reduction in the risk estimate, which is called **bias toward the null**. It has to be noted, however, that the presence of future cases in a control group does not create a false-positive finding. If an association is found, it is still valid, but the risk estimates might be a little lower.

Selection of cases is not straightforward

Not everybody with a disease is always a good choice for inclusion into a study. For almost any disease, it is necessary to include newly diagnosed **incident cases** into a case-control study. If existing patients diagnosed in the past (**prevalent cases**) are included, the sample will include a mixture of incident and prevalent cases. Depending on the disease, the prevalent cases may be a highly selected subgroup such as long-term survivors. In cancer research, if all patients seen in an outpatient clinic are enrolled in a study, patients with aggressive diseases who do not survive long after diagnosis will be underrepresented. **Figure 4.2** illustrates the difference between incident and prevalent case groups. In a study of HIV progression to AIDS, all cases should have known HIV-acquisition dates. Without this information, the case group will have a mixture of people: those living with HIV infection for a long time who have a short time left before developing symptoms of AIDS (assuming no treatment is given), and those who have recently become infected and, on average, will take longer to progress to AIDS. The outcome, time to AIDS, will be severely distorted if time-to-AIDS begins from the time of enrollment rather than the time of HIV acquisition. Also, the most severe cases will be missed in the prevalent subgroup.

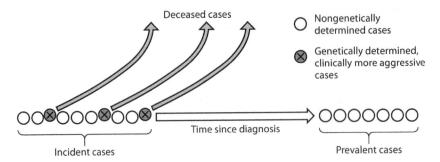

Figure 4.2 The difference between an incident case group and a prevalent case group. If all consecutively diagnosed cases are included in the study, this incident case group contains all genetically and nongenetically determined cases. Genetically determined cases tend to have clinically more aggressive disease and die earlier than nongenetically determined cases. If prevalent cases are recruited for a study, there will be a relative deficiency of genetically determined cases.

These problems associated with prevalent cases are called **survival time bias**. Thus, appropriate selection of cases is equally crucial as the selection of controls.

Comparability of cases and controls

The cases and controls must be similar in everything except the disease status. This means they have to be from the same population base so that they are exposed to similar environmental factors, and they must be similar in their age and sex distribution, as these are the most common variables that complicate epidemiologic research. There are also very good reasons to make sure that cases and controls have been selected contemporaneously in any association study, and particularly in genetic studies. In time, population composition may have changed due to immigration, and two samples recruited at different times from the same population base may be totally different genetically. It is a well-known issue for longevity studies in which centenarians are compared to younger people in the same geographic area, but these two populations may be inherently different. Such differences would yield a statistically significant difference in genotype frequencies that has nothing to do with the two groups being different in their case-control status. Besides genetic differences in noncontemporaneously sampled populations, environmental exposures may have been different too. It is, for example, recognized that *BRCA1* mutations have greater penetrance today compared with fifty years ago, presumably due to different environmental exposures today.

Cohort studies are prospective

Cohort studies take an opposite approach to case-control studies in that they start when everybody in the study is healthy and follow them until some of them develop the disease of interest (the outcome). Cohort studies therefore have a prospective design and begin with a defined group of individuals, such as everybody living in the same town (for example, the Framingham Heart Study) or everybody belonging to a profession or a society (for example, the Physicians' Health Study and the Nurses' Health Study). Since the study begins when every participant is healthy and all exposures are recorded as they occur, this design has a number of advantages over case-control studies (**Table 4.2**).

The main advantage of cohort studies is that exposure assessment is much more accurate than in case-control studies and does not rely on memory or recall. Cohort studies can also

Table 4.2 Advantages and disadvantages of cohort studies

Advantages	Disadvantages
Can establish population-based incidence	Takes long time to complete and requires a lot of resources
Accurate risk estimation	
Magnitude of a risk factor's effect can be quantified	May require very large samples
Allows repeated measurements	Not suitable for rare diseases or for diseases with a long latent period
Temporal relationship can be inferred (prospective design)	
Reverse causation is not a concern	Common bias types are nonresponse, migration, and loss to follow-up
Time-to-event analysis is possible	
Selection and information bias is less likely	Sampling, ascertainment, and observer biases and confounding are still possible
Multiple outcomes associated with the same exposure can be studied	
Causality is easier to establish	
Can be used where randomization is not possible	

establish the sequential relationship between exposure and outcome. Information on this sequence is important in establishment of causality, and also in ruling out a phenomenon called **reverse causation**. Sometimes the direction of an association is not clear: for example, cancer cases have low blood cholesterol levels and colon cancer cases have low iron levels. These associations are usually found in case-control studies and can be misinterpreted as low cholesterol and low iron increasing the risk of cancer. However, it is actually the outcome, cancer, that causes the decreased cholesterol (high metabolic rate in cancer) and iron levels (insidious bleeding in colon cancer). In a prospective design, the sequence of events is known and the correct temporal relationships can be worked out. For example, if cholesterol and iron levels are measured at the beginning of a cohort study, it will be clear that they decrease in those who develop cancer. Although cohort studies are able to establish the temporal sequence of exposure and outcome, and the reverse causation phenomenon is avoided, this feature is not relevant to genetic studies, as discussed later.

The main disadvantage of cohort studies is that they take a long time to reach the end point (usually many years) and following up so many people for so long requires a large amount of effort and resources.

Genotype is the exposure in genetic association studies

All epidemiologic studies deal with exposures (such as sunlight, smoking, or air pollution) and outcomes (such as cancer, asthma, or bronchitis). While traditional epidemiology focuses on environmental exposures, genetic epidemiology examines the associations between genetic factors and traits. A genotype, or the physiologic change caused by the genotype, is therefore the exposure in genetic epidemiology and is analogous with being exposed to an environmental factor. Since genotypes are present from birth, the exposure in a genetic association study (which is the genotype) always precedes the outcome. This is the reason why the reverse causation phenomenon is not a concern in these studies.

SNPs are **binary variables**. If one of the alleles (allele B) is associated with increased risk of disease, the other allele (allele A) or the genotype consisting of only allele A (AA) will be a marker for protection. In such binary variables, the risk changes associated with the two alleles will be reciprocals of each other. Thus, a twofold increase in risk (approximating to

Allele B-positive genotypes are the risk genotypes:

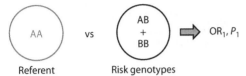

Referent   Risk genotypes

The association is converted to the opposite direction:

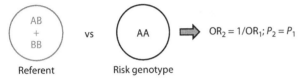

Referent   Risk genotype

Figure 4.3 Relationships between risk and protective associations by the same SNP. When allele B-positive genotypes show a risk association, allele A homozygosity shows a protective association. The statistical significance (measured by the P value) will be the same, and effect sizes (measured by the odds ratios) denoting the fold risk change will be reciprocals of each other.

odds ratio = 2.0) for one allele corresponds to a twofold decrease in risk (approximating to odds ratio = 0.5) for the other one, and these two risk changes are identical in magnitude.

By default, statistical analysis of associations is done assuming that the uncommon (minor) allele is the risk marker. Thus, genotypes that contain the minor allele (AB and BB) are treated as risk genotypes, and homozygosity for the major allele (AA) is treated as the baseline or referent. Depending on which genotype(s) is used as the referent, a risk association for one genotype may be a protective association for the other genotype of the same SNP. This is similar to smoking being a risk factor compared with nonsmoking, and nonsmoking being a protective factor compared with smoking for lung cancer. If both minor allele-positive genotypes (heterozygote AB and homozygote BB) equally contribute to increased risk, they are the risk genotypes, with AA being the referent. If (AB + BB) is used as the referent, AA will turn out to be a protective marker. When the risk change (estimated as an odds ratio) is converted to the opposite direction by taking its reciprocal, accompanying P values remain the same (**Figure 4.3**). Confusion may arise from a convention in epidemiology that an exposure showing an association is called a risk marker regardless of the direction of the association (risk or protection).

Genetic risk models resemble environmental risk models

Common genetic risk models share the same names as the Mendelian inheritance models discussed in Chapter 2, but do not apply to genotypes or alleles in the same way. In the **dominant model**, genotypes AB and BB are equally associated with risk, and if genotype BB is the only risk-associated genotype, it is the **recessive model**. As **Table 4.3** shows, dominant and recessive risk models are analogous to the presence or absence of exposure to an environmental factor. The analysis of genetic risk models is also similar to the analysis of an environmental variable that indicates the presence or absence of an exposure in

Table 4.3 Common genetic risk models and analogies with environmental risk models

Genetic risk model	Referent genotype(s)	Risk genotype(s)	Analogy
Dominant model	AA	AB + BB	Absence or presence of an exposure
Recessive model	AA + AB	BB	Absence or presence of an exposure
Additive model	AA	AB < BB	Dosage effect

nongenetic studies, such as smoking or not smoking. In epidemiologic association studies, the dosage of exposure is equally as important as the presence or absence of an exposure.

Genotypes are analyzed for a gene dosage effect using the **additive model**. This model may be likened to the examination of association for different grades of an exposure, such as no smoking, moderate smoking, and heavy smoking. In the additive model, all three genotypes are retained as individual exposures and represent gradually increasing levels of exposure from AA (baseline), to AB (intermediate exposure), and to BB (maximum exposure).

Family-based study designs offer a good alternative to the case-control design

Case-control studies are the most common type of study in genetic association studies, but cohort studies, which require greater resources, yield more reliable and robust results. In genetic association studies, a major complication stems from having multiple subsets within a population, called population substructure; this can be alleviated by using a **family-based design**. The population substructure concept and how it confounds case-control studies is discussed in Chapter 3. In the most common form of a family-based study, the parents and affected child form a triad and these triads are enrolled in the study. The information sought in each triad is whether the affected child inherits any particular allele from a heterozygote parent more than 50% of the time. This possibility is tested by the **transmission disequilibrium test** (**TDT**). Homozygote parents do not provide any information for this analysis since they have two copies of the same allele and all offspring receive the same allele. It is, therefore, not possible to assign which allele has been transmitted to the affected offspring. The main advantage of the family-based design, in general, is its robustness for population substructure. While case-control studies are extremely sensitive to this possibility, family-based study results are not because no comparison is made with anyone who may have a different genetic background.

A family-based study can also be turned into a case-control study. A triad allows determination of which two of the four parental alleles are inherited by the affected child, and the other two alleles may be assigned to a pseudo-sibling. Thus, there will be as many pseudo-siblings as cases in the study; this results in a special type of case-control study in which each case is matched to a pseudo-control (**Figure 4.4**). In the first example given in Figure 4.4, the affected child has inherited allele B from each parent, and the matched pseudo-sibling will have the genotype AA consisting of the two parental alleles not inherited by the affected child. If there is more than one unaffected sibling, they all can be used as pseudo-sibling controls.

One obvious limitation with the family-based design is the need for biological material from both parents, which is not always possible, especially for studies of late-onset diseases. This design is therefore most commonly used for childhood diseases. Compared with case-control studies, triads are not only more difficult to recruit but also more costly. Instead of a case-control pair, three samples from the family triad need to be genotyped,

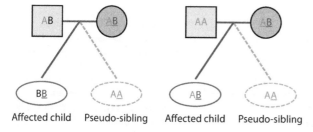

Figure 4.4 Pseudo-sibling design using data generated in family triads. Affected children's genotypes will consist of alleles inherited from parents, and pseudo-siblings have the alleles that are not transmitted to affected children.

and homozygote parents cannot be used. As homozygote parents do not contribute to the analysis, the number of triads to recruit may also have to be increased for the study to have sufficient statistical power. Family-based studies are superior to case-control studies in ethnically/racially diverse populations and for childhood diseases where both parents are involved in the clinical workup, such as in preparation for hematopoietic stem cell transplantation. The best use of this design is in the replication arm of a case-control study to rule out population substructure as an alternative explanation for the association noted in the initial case-control study.

Family-based designs do not always rely on parents and affected child triads. Alternative designs have been developed that use only siblings (sib-TDT study) or more distant relatives. Such models overcome the drawback of triad-based designs by being suitable for use in late-onset diseases for which parental genotypes may be difficult to obtain. The sib-TDT model uses affected and unaffected sibling genotypes that are obtained more easily than those of parents.

4.2 Chance, Bias, Confounding, and Effect Modification

In any epidemiologic research, **chance**, **bias**, and **confounding** are elements that threaten the validity of the study by distorting the results (**Figure 4.5**). Distortion can occur in two ways: it can cause false negatives or false positives. It seems to be customary that when a positive result is obtained, alternative explanations are explored, but when the result is negative, the same is not tried as intensively. This is probably a result of the perception that false positives are a more important issue than false negatives, but it is only fair to treat false negatives and false positives as equally wrong results, and to consider alternative explanations at the end of every study.

While bias is a systematic kind of deviation from the truth, chance is a random deviation. Variability through chance can occur despite perfect study design, execution, and analysis. When the sample size is large enough, errors caused by chance will be canceled out, but errors from bias will remain whatever the sample size. As a result, chance leads to imprecise results, and bias leads to invalid results. The best way to avoid chance is to have a large sample size. Avoidance of bias requires much more than that, beginning from the design phase of a study.

There are many types of bias

Systematic deviation from the truth can occur for many reasons. Bias usually results from recognizable errors in study design or conduct of the design. Unlike confounding, as discussed later, the consequences of bias cannot be fixed at the analysis phase. Bias may result in a weakening or underestimation of a true association, or may create a false

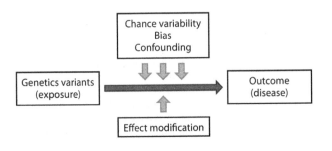

Figure 4.5 Modifiers of the exposure–disease relationship in association studies. Chance, bias, and confounding may cause some deviation of the results from true results. When effect modification is considered, it may help to unmask otherwise undetected results.

association. Which one will occur depends on whether the bias was **nondifferential** (applies to both comparison groups) or **differential** (applies to one of the comparison groups).

There are a large number of bias types. The two most common bias categories are selection and information biases. Any design error that violates the representativeness and randomness of study subjects is a **selection bias**. It occurs when the opportunity to be included in the study population is dependent on the outcome or exposure of the study. The most common examples of selection bias include usage of convenience samples as controls, mixing incident and prevalent cases in the case group, and low participation rates. The **information bias** category includes errors in measurements of exposures or outcome variables that provide wrong information for analysis. Information bias most commonly occurs due to **recall bias** in case-control studies because information is collected retrospectively. Recall bias is a differential bias due to cases and controls remembering their past exposures differently and generating a nonexistent difference, which in turn results in a spurious association. All DNA extractions, genotypings, and genotype callings should be performed in the same way for all samples. Otherwise, information bias may result from obtaining different results due to differences in the quality of DNA or the methods used to genotype individuals and to assign genotypes, not real differences.

Confounding is possible in genetic association studies

Confounding is sometimes referred to as a form of bias and called confounding bias. Confounding should be seen as a problem of interpretation. If a result is confounded, the observed relationship between the risk factor/exposure and the disease is not causal (or partially not causal) but is explained by their joint correlation with the true but unmeasured independent risk factor/exposure (**Figure 4.6A**). Confounding can only occur if the prevalence of the confounding factor (or confounder) differs between groups being compared. A confounder (true risk factor) is not on the **intermediate pathway** between the exposure examined and the outcome.

Besides the universal confounders in any epidemiologic study, such as age and sex, ethnicity is an important factor to consider in genetic association studies. A major confounder in genetic association studies is where the unknown **causal variant**(s) is in strong linkage disequilibrium (LD) with the noncausal variants being examined (linkage disequilibrium is discussed in Chapter 3). This situation is called **confounding by locus** (**Figure 4.6B**) and applies to all indirect associations. It is the causal variant that is actually responsible for the association signal, but the association is attributed to the examined variant that is correlated with the causal one. To identify the causal association, all variants in LD with the associated marker may have to be examined. This is usually achieved by sequencing of the region to make sure that even the as-yet unknown variants have been assessed. The causal variant yields the strongest signal and shows some functionality.

In essence, a confounded association is not an association with the true risk factor but with a surrogate that correlates with the true risk factor. The causal factor and the surrogate marker correlated with it in the study are said to be confounded when their separate causal effects are not clear. While no statistical association is necessarily causal, a confounded association is at best a partially causal one and identifies a risk marker and not the true risk factor. The correlation between the causal factor and its surrogate factor may be complete, such that the observed association is totally due to the correlation (in which case, the observed association disappears when statistically adjusted for the causal factor).

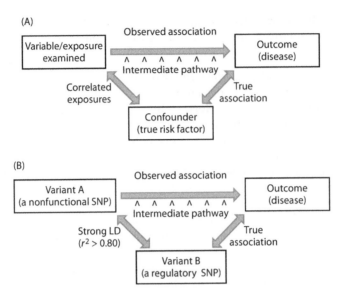

Figure 4.6 Confounding in nongenetic and genetic studies. (A) Relationships among the observed association, a confounder, and outcome in association studies; (B) confounding by locus in genetic association studies. In a confounded association, the observed association is totally or partially due to the correlation between the associated variant and the causal variant.

The correlation may also be incomplete, in which case the statistical adjustment only weakens the observed association, suggesting that the surrogate marker may still have some independent association and partial causality. If an association is a confounded one for which there is no statistical test to assess it, the associated marker will not pass the causality assessment, and any intervention that is based on this finding will not work. For example, potential associations that can be found between coffee-drinking or alcohol consumption and lung cancer are not true associations, but confounded ones; they come about because of their correlation with the true risk factor (confounder), which is smoking.

Confounders can be positive or negative

Confounders may cause overestimation of an association (**positive confounding**) or underestimation of an association (**negative confounding**). The overestimation of an association inflates the effect size and moves it away from the null association value (odds ratio of 1.0). Negative confounding shifts the effect size toward the null association value. A positive confounder is correlated with exposure and outcome in the same direction (a causal variant in LD with the examined variant); a negative confounder shows an inverse relationship with the associated variant (a causal variant in negative LD or inversely correlated with the examined variant). Statistical correction of results involving a positive confounder will result in a decrease in the magnitude of the effect size by correcting the overestimation, but adjustment for a negative confounder will increase effect size by correcting its underestimation.

A frequently cited example of negative confounding is the role of age in the association between oral contraceptive (OC) usage and myocardial infarction (MI). Age is a risk factor for MI and is inversely correlated with OC usage (the higher the age, the less OC usage). In this case, adjustment for age will increase the OR because of negative confounding by age. Another example of negative confounding is the association of smoking during pregnancy and low birth weight of the offspring. Both smoking and advanced age are associated with lower birth weight, and smoking in pregnancy is inversely correlated with age. The association of smoking with low birth weight gets stronger when adjusted for age because of the inverse correlation of smoking with age and also that they both have effects on birth weight. There are examples in the literature where the adjustment for maternal age requires the use of exact age in years rather than broad categories (like below 36 and above 35 years). If broad categories are used, the whole of confounding may not be controlled and some **residual confounding** may remain, resulting in a partially confounded association of smoking with birth weight.

In positive confounding, the observed association either disappears or gets weaker when adjusted for the confounder. An example is the association of Down syndrome with birth order that is confounded by maternal age, which is positively correlated with birth order (mothers of offspring with higher birth order are also older). Thus, birth order acts as a surrogate for the actual risk factor (maternal age). A study examining the association of birth order would find an association between higher birth order and Down syndrome, but this association would disappear when corrected for maternal age. In this example, it is the relationship between birth order (examined variable) and maternal age (true risk factor) that confounds the association of birth order with Down syndrome.

Confounding may mask associations

Confounding should be considered not only when an association is found, but also when an expected association is not observed, as it may have been masked by negative confounding. In genetic association studies, it is possible that population substructure (see Chapter 3) may mask an existing association if the effect of the stratification is in the opposite direction to the association. Confounding may even change the direction of the observed effect. However, it should not be assumed that all observed associations are confounded or that all negative findings are due to negative confounding.

One of the threats to the validity of a case-control study is the unintentional differences between cases and controls that may confound the observed associations. Every effort should be made to select cases and controls from the same study base at the same time period. To assure that this has been achieved, as many variables as possible should be examined in cases and controls, and a table comparing them and presenting summary statistics such as frequencies, means, or medians of all potential confounders should be presented at the beginning of the results section of any report. No remarkable differences between cases and controls for examined potential confounders (age, sex, ethnicity, education levels, socioeconomic level, and so on) will reassure that cases and controls are comparable and indicate that residual confounding is unlikely.

Confounding may be controlled at the design and analysis phase of a study

When confounding is suspected and data have been collected on potential confounders, there are ways to explore confounding and to identify the relationship between variables that confounds the observed association. If ethnicity is a suspected confounder, the study or the analysis may be restricted to one ethnicity only (**restriction**; also called **stratification** if the sample is divided in the analysis phase). Since a variable can only be a confounder if its frequency is different between study groups, restriction to one category of a confounder makes sure that the groups cannot differ in its frequency. Thus, in a study that uses a multi-ethnic sample, an observed association in the whole sample may disappear if the study is restricted to one ethnicity (restriction) or if the analysis is done by ethnicity (stratification). In a study restricted to one ethnicity or where the analysis is done for each ethnicity separately, there will be no association if the original association in the multi-ethnic sample was confounded by the relationship of the associated marker and one of the ethnicities. In the Down syndrome example, the birth-order association will disappear if the sample is divided in age groups (stratification) and analyzed for each age group separately. In addition, in each birth-order category, there will be an association between maternal age and Down syndrome risk.

A similar method is **matching**, but this is applied in the design phase where cases and controls, or subjects in comparison groups, are matched for potential confounders. This can be one-to-one matching (for each case from a certain ethnicity, for example, one or more control subjects from the same ethnicity are included) or frequency matching (if half the cases are from a certain ethnicity, around half of the controls are included from the same ethnicity, and so on). This way, the frequency of the potential confounder (ethnicity for this example), is not different in the two groups, and confounding by the matched factor cannot occur. If a matched design is used, an appropriate statistical test should be used for matched analysis. A major disadvantage of matching is that the effects of the matching variable cannot be examined in that study. In a study that uses sex-matched cases and controls, for example, sex cannot be a confounder, but any sex effect cannot be examined either.

Statistical adjustment can be applied at the analysis phase. If an association changes during the analysis of results when a potential confounder (for example, age) is added to the statistical model, it is an indication of confounding. In a case-control study, the crude OR for the observed association changes when the results are corrected for the potential confounder (for example, age). As a rule of thumb, the corrected or adjusted OR should be approximately 10% different from the crude OR to conclude that confounding exists. The availability of methods to adjust for confounders means that data should be collected for all potential confounders, which are usually all known risk factors, whenever possible.

Box 4.1 presents a common example of confounding by ethnicity in a genetic association study, and shows how the correct results can be obtained.

Unknown confounders may be controlled too

When potential confounders are not known, it is not possible to collect data for them. Under these circumstances, only a randomized study design can be used to control confounding, which is not possible in a case-control study. **Randomization** assures equal distribution of unknown confounders between the groups. Randomized studies have limited use in genetic association studies. One reason is that unlike traditional epidemiologic studies, genetic association studies are not equally confounded by environmental and behavioral factors as they do not strongly correlate with genotypes. Major confounders in genetic association studies are ethnicity and causal genetic variants in LD, and there are easier ways to control for these. One use of randomized clinical trial design is the testing of a genotype-by-drug interaction when preliminary studies suggest differential drug effects by genotype. Such an interventional study has very limited value for examination of the genetic basis of disease susceptibility.

Intermediate variables should not be controlled for

Since a variable on the intermediate pathway between exposure and outcome is not a confounder, they should not be used to adjust associations. For example, in a study of

---

Box 4.1 Methods for correcting confounding by ethnicity

A genetic association study uses a multi-ethnic sample and finds a genetic association with a SNP that is unlinked to the disease but has a higher frequency among Europeans (it is an ancestry informative marker). The disease is also more common among Europeans so there are more Europeans in the case group than the control group. Thus, by virtue of having a higher proportion of Europeans among cases, a SNP that is more common in Europeans appears to have an association with the disease that is confounded by the relationship of this SNP to European ethnicity (**Figure 1**).

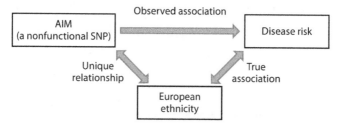

Figure 1 A confounded association.

This result can be assessed for confounding by ethnicity by any of the following methods:

- Restriction: if the study is conducted in Europeans only, the SNP frequency will not differ between cases and controls (all Europeans) → no association

- Stratification: if the multi-ethnic sample is analyzed after dividing the data into subsets for each ethnicity, the SNP frequency will not differ between cases and controls in any ethnicity → no association

- Matching: if each European case is matched to a European control (and each of the other ethnicities are also matched similarly), there will be no frequency difference for the SNP in matched analysis → no association

- Statistical adjustment: if the data from the multi-ethnic sample are corrected for ethnicity, the initially observed association will disappear → no association

---

correlation between overeating and obesity, blood sugar or lipid levels are not confounders, and adjustments by them would not be appropriate. Such an adjustment would result in masking of a true association between overeating and obesity. Inclusion of known risk factors in the analysis of an association with disease risk is highly recommended. However, in a genetic study, nongenetic factors known to be associated with the risk may be in the intermediate pathways between the genetic risk marker and disease development. Thus, inclusion of such nongenetic risk factors in the analysis of genetic association results requires attention. If a genetic marker reduces cardiovascular disease risk via its favorable effects on blood lipids, inclusion of measured blood cholesterol levels as a known risk factor in the analysis will diminish the strength of the genetic association.

Effect modification is not a bias and provides useful information

Unlike bias and confounding, effect modification is a useful feature that provides information on the nature of an association. It does not violate the internal validity of the study and has nothing to do with sample size or chance. If effect modification is present, two subgroups (such as males versus females, early-onset versus late-onset disease, or Europeans versus non-Europeans) will show different association statistics. It can be checked because of prior knowledge or explored by use of a **statistical interaction** test. Such tests (explained in more detail in Chapter 6) examine whether an association differs when analyzed in two different subsets of the sample (for example, in males and females). If an

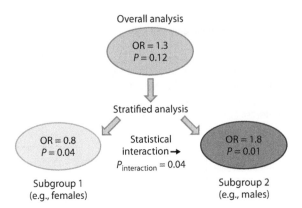

Figure 4.7 Effect modification. A negative finding in the overall analysis of association may be due to modification of the effect by a factor splitting the sample into subgroups. In this example, there is no statistically significant association in overall analysis. In each subgroup, however, there are statistically significant associations in different directions, resulting in a statistical interaction. The statistically significant interaction indicates that the difference between the associations observed in the two subgroups is statistically significant. The interaction may result from associations in different directions, as in this example, or from associations differing in strength as measured by the odds ratio.

interaction is indicated, stratified analysis follows for different levels of the interacting factor (such as sex) and yields different ORs in each stratum of the effect modifier. The difference in ORs may be simply in strength (for example, 1.8 and 3.2) or also in direction (for example, 0.8 and 1.8) (**Figure 4.7**).

Statistical adjustment in analysis does not assess effect modification. Such statistical adjustment removes confounding but does nothing about effect modification. Unraveling an effect modification helps with the interpretation of the association and may shed light on the mechanism of the association and the etiology of disease. For example, finding out that the association is restricted to males could suggest that sex hormones or sex chromosomes may be involved in disease etiology. Exploration of effect modifications is an underused tool in genetic association data analysis. One reason is the misperception that statistical adjustment rules out effect modification. Effect modifications should be sought by interaction or stratified analyses as vigorously as confounding is explored.

Table 4.4 summarizes the key features of chance, bias, confounding, and effect modification in association studies.

Table 4.4 Key features of chance, bias, confounding, and effect modification

	Chance	Bias	Confounding	Effect modification
Threat to validity	Yes	Yes	Yes	No
Can generate a spurious result	Yes	Yes	Yes	No
Can mask an existing association	Yes	Yes	Yes	Yes
Correlates with sample size	Yes	No	No	No
Corrected by statistical adjustment	No	No	Yes	No
Can be identified by stratified analysis	No	No	Yes	Yes

4.3 Causality and Statistical Association

A statistical association is just that. Observing a correlation between two variables does not suggest one of them causes the other. In general, an association between two variables may be causal if it is strong, consistent, specific, and plausible, follows a logical time sequence, and shows a dose–response gradient. This principle derives from **Hill's criteria for causality (Box 4.2)** and applies equally to all epidemiologic studies. The identification of confounding does not necessarily suggest causality by the confounding variable. The association of the confounding variable is more likely to be causal than the confounded association, but it is still a statistical association and requires careful examination of causality.

The logical time sequence or temporality is a given in genetic association studies, since germ-line polymorphisms (exposure) are present at birth and always precede the outcome. The dose–response gradient is assessed by the association under an additive genetic model. An additive model association supports causality, but its absence does not rule it out. Plausibility is usually a problematic issue. It is not hard to find some evidence for the biological plausibility of any given genetic association with a phenotype, but this has to be based on a prior hypothesis and should not be driven by data. In genetic association studies, an observed association is usually an indirect one, with a marker in LD with the unknown causal disease marker. One of the most important indicators of causality in a genetic association study is the functionality of the associated marker, which is considered under "plausibility" and "experimental confirmation" (see Box 4.2). If the marker has no

---

Box 4.2 Hill's criteria for causality and their application to genetic association studies

- Temporal relationship between exposure and outcome: the exposure should precede the outcome so there is no reverse causation. This is always assured for a genetic association as genotypes (exposure) are present from conception and always precede the outcome (disease).

- Strength of the association: the stronger a statistical correlation, the more likely that it is causal. This is measured by the effect size and not by statistical significance.

- Biological gradient (dose–response relationship): correlation between the dose of the exposure and the change in the risk suggests a causal relationship. This is assessed by the additive genetic model in a genetic association, but its absence due to strictly recessive or dominant genetic associations does not rule out causality.

- Consistency: replication of the results always increases the validity of results and makes causality more likely.

- Plausibility: biological plausibility of a genetic association increases the likelihood of it being a causal association. For greater credibility, plausibility should be hypothesized before the study.

- Alternative explanations: ruling out alternative explanations such as chance, bias, and confounding increases the likelihood of causality.

- Specificity: if the outcome cannot be attributed to any other exposure, causality is more likely. This is more important in nongenetic associations as the same gene (variant) may be causally associated with more than one disease, and multiple variants are expected to play a role in the pathogenesis of common diseases.

- Coherence: the biological plausibility of the association should not conflict with the generally known facts of the biology of the disease.

- Experimental confirmation: functional replication of an association.

- Analogy: the effect of similar exposures should be similar on the outcome. Different genetic variants affecting the same gene should yield similar results if they are causal.

known function, the probability of causality is reduced, whereas a highly functional polymorphism is more likely to be the causal marker. Functionality of a genetic variant is assessed experimentally or by bioinformatics, which will be discussed in Chapter 10.

Key Points

- An epidemiologic association provides a statistical result, and an association does not mean causality.
- Cases and controls need to be selected carefully, as poor selection of either can invalidate results.
- Chance, bias, and confounding can distort results either way: they can cause false positives and false negatives.
- Confounding is not an all-or-none phenomenon. The results may be partially contributed to by a confounder. If statistical adjustment for the potential confounder changes the result, it is a biased result without the adjustment.
- Unfortunately, most confounders are unknown and the only way to have a nonconfounded result is to do a randomized study, which is not possible for observational genetic association studies.
- To a degree, genotypes can be seen as no different from any other variables. In the analysis of single SNP associations, the analysis is not different from examining an association with the presence or absence of an environmental exposure or its dosage.
- Technologic advances in genotyping efforts have helped with minimizing measurement error but are no replacement for methodologic safeguards. Principles of epidemiologic research need to be adhered to regardless of the technology used.
- The epidemiologic knowledge should be applied not only to conduct a perfect study, but also to assess the imperfections that may have impacted the results. Each epidemiologic study report should include a self-critical assessment of any remaining bias that may have caused deviation of the results from the truth.
- Observational study results should be interpreted after excluding any selection bias, information bias, confounding, and chance. Any result with internal validity should then be replicated and functionally assessed for causality.

Further Reading

General epidemiology

Bhopal R (2009) Seven mistakes and potential solutions in epidemiology, including a call for a World Council of Epidemiology and Causality. *Emerg Themes Epidemiol* 6, 6 (doi: 10.1186/1742-7622-6-6). (*The most common errors in epidemiologic studies are discussed with examples and their solutions are given. Included among the errors are "insufficient attention to evaluation of error" and "either overstatement or understatement of the case for causality."*)

Blair A, Saracci R, Vineis P et al. (2009) Epidemiology, public health, and the rhetoric of false positives. *Environ Health Perspect* 117, 1809–1813 (doi: 10.1289/ehp.0901194). (*The papers by Blair et al. and Boffetta [below] point out the weaknesses of observational studies and, by doing so, they provide a good review of chance, confounding, and bias with examples from the literature. Both false positives and false negatives are discussed.*)

Boffetta P (2000) Molecular epidemiology. *J Intern Med* 248, 447–454 (doi: 10.1111/j.1365-2796.2000.00777.x).

Boffetta P, McLaughlin JK, La Vecchia C et al. (2008) False-positive results in cancer epidemiology: a plea for epistemological modesty. *J Natl Cancer Inst* 100, 988–995 (doi: 10.1093/jnci/djn191).

Bonita R, Beaglehole R & Kjellström T (2006) Basic Epidemiology, 2nd ed. World Health Organization. (*This edition is accessible via the WHO Website: http://apps.who.int/iris/bitstream/10665/43541/1/9241547073_eng.pdf.*)

Burton PR, Tobin MD & Hopper JL (2005) Key concepts in genetic epidemiology. *Lancet* 366, 941–951 (doi: 10.1016/S0140-6736(05)67322-9). (*First paper in the Lancet septet on genetic epidemiology. It contains useful information on the approach to identify genetic determinants of complex disease, and how genetics and epidemiology are fused to achieve this aim.*)

Coggon D, Rose G & Barker DJ (1997) Epidemiology for the Uninitiated, 4th ed. BMJ Books. (*This edition is accessible via the BMJ Website: http://www.bmj.com/about-bmj/resources-readers/publications/epidemiology-uninitiated.*)

Silman AJ & Macfarlane GJ (2002) Epidemiological Studies: A Practical Guide. Cambridge University Press. (*Described very aptly by its title, as "it explains the principles and practice of epidemiology and serves as a handbook for those who wish to do epidemiology." It is full of real-life examples to prepare the reader in how to set up a study and produce valid results. This book connects theory with practice in the best possible way.*)

Epidemiologic study designs

Cordell HJ & Clayton DG (2005) Genetic association studies. *Lancet* 366, 1121–1131 (doi: 10.1016/S0140-6736(05)67424-7). (*Another paper from the Lancet septet, which reviews study designs in the context of genetic association studies.*)

Gallacher J (2009) Epidemiologic considerations in complex disease genetics. In Genetics of Complex Human Diseases (Al-Chalabi & Almasy eds). *Cold Spring Harb Protoc* 2012 (doi: 10.1101/pdb.top067512).

Hattersley AT & McCarthy MI (2005) What makes a good genetic association study? *Lancet* 366, 1315–1323 (doi: 10.1016/S0140-6736(05)67531-9). (*This fifth paper in the Lancet septet discusses with clarity the epidemiologic, statistical, and methodologic issues relevant to genetic association studies.*)

Zondervan KT & Cardon LR (2007) Designing candidate gene and genome-wide case-control association studies. *Nat Protoc* 2, 2492–2501 (doi: 10.1038/nprot.2007.366). (*Reviews the most common study design in genetic association studies and provides a step-by-step protocol.*)

Bias and confounding

Grimes DA & Schulz KF (2002) Bias and causal associations in observational research. *Lancet* 359, 248–252 (doi: 10.1016/S0140-6736(02)07451-2).

McNamee R (2003) Confounding and confounders. *Occup Environ Med* 60, 227–234 (doi: 10.1136/oem.60.3.227).

Okasha M (2001) Interpreting epidemiological findings. *Student BMJ* 9, 305–306 (http://student.bmj.com/student/view-article.html?id=sbmj0109324). (*In just two pages, this paper provides a very concise guide to the language of epidemiology and alternative explanations to epidemiologic study results.*)

Smith GD & Ebrahim S (2002) Data dredging, bias, or confounding. *BMJ* 325, 1437–1438 (doi: 10.1136/bmj.325.7378.1437).

Vineis P & McMichael AJ (1998) Bias and confounding in molecular epidemiological studies: special considerations. *Carcinogenesis* 19, 2063–2067 (doi: 10.1093/carcin/19.12.2063).

Causality assessment

Hill AB (1965) The environment and disease: association or causation? *Proc R Soc Med* 58, 295–300. (*The original paper that established the causality criteria known as Hill's criteria for traditional epidemiologic studies.*)

Page GP, George V, Go RC et al. (2003) "Are we there yet?" Deciding when one has demonstrated specific genetic causation in complex diseases and quantitative traits. *Am J Hum Genet* 73, 711–719 (doi: 10.1086/378900). (*Building on Hill's criteria, this paper discusses the criteria for declaring a genetic association to be with the causal variant.*)

Environmental Epidemiology in the Context of Genetic Epidemiology

5

Traditional epidemiology is concerned with environmental causes of disease and can be thought of as environmental epidemiology. Genetic epidemiology investigates genetic causes of disease, with the new subdiscipline of epigenetic epidemiology acting as an interface between environmental and genetic epidemiology. The false dichotomy of nature versus nurture presents the effects of the environment and genes as mutually exclusive, but this is an unproductive approach. There are only a few strictly environmental or genetic conditions, and most diseases are contributed to by both environmental and genetic risk factors.

A phenotype is the sum of the contributions from environmental and genetic factors. Even if a single environmental or genetic factor does not show an effect, they may jointly affect the phenotype. Recent advances in human genetics may have created an impression that genetics has the upper hand compared with the environment in its contribution to phenotypes, but this is not always the case. This chapter will consider the role played by the environment and the interplay between genetics and the environment in disease causation.

The environment and genotype do not act in isolation

The continuum of disease phenotypes has at one end those diseases determined by fully penetrant mutations and, at the opposite end, those determined purely by environmental effects. Those disorders resulting from incompletely penetrant genetic effects and multifactorial (complex) disorders lie in between (**Figure 5.1**). Even within the group of single-gene disorders, there is a continuum from purely genetic to mixed etiology. Fully penetrant mutations are those that abolish the activity of a gene altogether, with deleterious consequences, without needing any contribution from another gene or environmental agent.

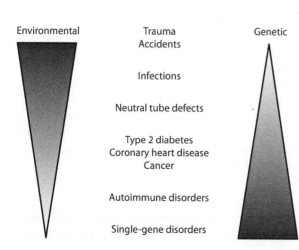

Environmental

Trauma
Accidents

Infections

Neutral tube defects

Type 2 diabetes
Coronary heart disease
Cancer

Autoimmune disorders

Single-gene disorders

Genetic

Figure 5.1 Contribution of environmental and genetic risk factors to the development of various conditions. Trauma and accidents have almost exclusively environmental etiology, whereas single-gene disorders are mainly genetic.

For example, on their own, mutations in the *TP53* gene cause Li-Fraumeni syndrome and those in the *F8* gene cause hemophilia A. A few single-gene diseases placed at this end of the spectrum are not caused purely by highly penetrant mutations. Phenylketonuria (PKU), for example, is an autosomal recessive disorder but, in the absence of phenylalanine intake, the mutation itself does not cause any problems. Likewise, hemolytic anemia due to glucose 6-phosphate dehydrogenase (G6PD) deficiency occurs only after ingestion of fava beans or sulfonamides. The penetrance of *BRCA1* mutations that cause breast cancer and *HFE* mutations that cause hereditary hemochromatosis is also modified by environmental factors as well as other genetic factors.

5.1 Studying Environmental Effects

The environmental contribution to disease susceptibility is studied by traditional epidemiology in association studies, but there is no equivalent yet of hypothesis-free **genome-wide association studies** for environmental factors. However, it is possible to estimate the relative contribution of the environment in studies actually designed to detect genetic contributions. When no evidence for a genetic contribution is found, it may be concluded that the environment plays a greater role than genetics. The following findings suggest a predominant environmental over genetic contribution to etiology:

- No difference in disease concordance rates between dizygotic and monozygotic twins
- Increases in disease frequency in successive generations of migrants compared with the frequency in their native land
- Risk in spouses of cases exceeding the risk in blood relatives of cases
- No change in disease frequency in inbreeding studies
- Lack of familial clustering
- Lack of genetic associations with a disease despite multiple well-powered studies in multiple populations

The genetic versus the environmental etiology of cancer is well studied

Cancer provides a good example for understanding the joint contributions of the environment and genetics to disease. Cancer has been extensively studied because of the presence of certain cancer subtypes that are purely genetically determined. Cancer is often described as a genetic disorder but, with the exception of rare hereditary cancer types, the genetic changes in cancer cells are acquired after neoplastic transformation and are therefore **somatic mutations**. Thus, not all cancers are due to inherited genetic mutations, even though somatic genetic changes in cancer cells are always present. It is not plausible that the recent increase in cancer incidence is due to sudden changes in genotype frequencies; changes in patterns of exposure to environmental factors are the more likely cause. It is, for example, known that the penetrance of *BRCA1* mutations has increased over a number of decades due to the increasing intensity of environmental modifiers of breast cancer risk. Obviously, the frequency of *BRCA1* mutations has not changed in such a short time, but a higher proportion of mutation carriers now develop breast cancer compared with 50 years ago.

Cumulative data mainly from familial cancer cases and twin studies suggest that environmental etiology is stronger than heritable etiology in cancer. One parameter that quantifies the relative contribution of each component in disease development is the **population attributable fraction** (PAF). PAF is the expected proportional reduction in the disease incidence in a population if exposure to a risk factor could be reduced to an ideal level, such as the reduction in lung cancer cases if all smokers quit smoking. For environmental factors, PAF is estimated to be up to 90% in cancer overall. This leads to an interesting conclusion: if Western populations were to live in the same conditions as the populations of developing countries, the risk of cancer would decrease by 90%, provided that the major carcinogen exposures—viral infections and mycotoxins—are avoided. Not everyone is equally susceptible to environmental risk factors and the effect is modified by genetic variation, but if there is no environmental exposure, genetic constitution is irrelevant.

The stronger role played by environmental factors in cancer causation has a major implication for the way in which genetic effects are examined. Most genetic markers of cancer susceptibility are likely to be modifiers of environmental effects rather than initiators of cancer development. Another implication is that when a genetic association is found, the observed main effect incorporates the average effects of environmental factors, even though the exposure and who has been exposed are unknown. If genetic association studies could be restricted to people who have been exposed to known environmental factors, they would yield stronger associations. Since environmental exposures are variable across populations, it is only natural that the strength of the same genetic association varies among populations.

The interaction between the environment and genetics has implications for the interpretation of genetic association studies

Previously found genetic associations are often not observed in a second study, which is taken to be a **replication failure**. If the first study was not conducted properly and its finding was spurious, it will never be replicated. There are good reasons, however, for failing to replicate a genuine association. Environmental variability is one reason. A hypothetical example of this phenomenon is given in **Figure 5.2**. As an example of a nonlinear correlation, both iron deficiency and iron excess increase the risk for cancer development. In an iron-deprived population, a polymorphism correlating with body iron levels will have a protective association with cancer. In an iron-replete population, the same polymorphism will appear to correlate with increased risk. Similarly, the correlation between the *MTHFR* variant 677 C > T and blood homocysteine concentration shows variation depending on the environment. The *MTHFR* variant shows a stronger correlation with homocysteine levels in low-folate regions than in areas with dietary folate fortification.

A complex disease is caused by interplay among multiple genetic variants and environmental factors. In this highly complex scenario, a genetic association study is trying to find one of the genetic factors. A factor may show an association in one study, due to its high frequency and its unknown but favorable genetic interactions, but not in a second study where these unknown interactions may not occur in a different genetic background. Like the hypothetical case of opposite genetic associations in different environments presented in Figure 5.2, the genetic association may also be reversed in a genetically different population. This is called a **genetic flip-flop phenomenon**. In purely genetic associations, the flip-flop phenomenon may occur due to differences in LD patterns between the causal and observed genetic variants in different populations. **Box 5.1** lists some possible scenarios of where a previously found, robust association is not observed in a well-designed study in a different population.

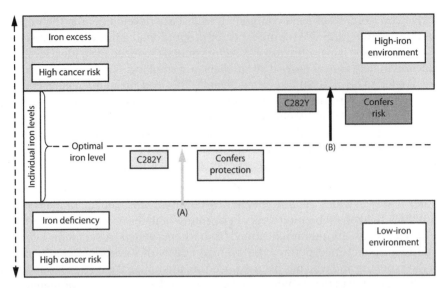

Figure 5.2 The effect of a genetic variant may be different depending on the environment.
The same variant in an iron regulatory gene that increases body iron levels may show opposite associations with cancer in different populations that have variable average iron levels; this is because both low and high iron levels are risk factors for cancer susceptibility. In an iron-depleted population (A), C282Y increases the iron levels from iron-deficiency levels to optimal levels and confers protection. In an iron-replete population (B), the same mutation elevates body iron levels to iron-excess levels and confers increased risk for cancer.

Box 5.1 Plausible genetic and environmental reasons for the lack of an association signal in a second study

- If the first association was an indirect one (not with the causal variant, but confounded), then it was a marker that is in linkage disequilibrium (LD) with the unknown causal variant. LD may be different or even nonexistent in a different population. The first, indirect association will not be observed in a second study.

- If the first association was due to an unknown gene through gene interaction (epistasis) in the first population, but the interacting gene variant exists at low frequency in another population, the association may not be observed in the second population.

- Some genetic associations arise due to modification of unknown environmental effects by genotypes. If the unknown environmental exposure differs in frequency or magnitude between populations, there will be no consistency between associations.

- Environmental differences can result in a change in the direction of an association. For a nutritional element that shows a nonlinear correlation between its level and disease susceptibility, if the level or bioavailability of that element is altered by a genetic variant, the associations may differ between populations where those populations differ in average levels of the element.

5.2 Examining and Estimating Environmental Effects

Genotyping is a straightforward process with very high reproducibility, but measurement of an environmental exposure is not an easy task. Exposure to most environmental factors shows temporal and spatial variation. Thus, the timing, amount, duration, and intensity of exposure have to be recorded to obtain an accurate measurement of environmental exposure and to avoid misclassification of exposure. Episodic exposure, exposure to complex mixtures, low-level exposure that may have a cumulative effect, and (in the case of biomarkers of exposure) a short half-life of a biomarker also make environmental exposure measurement much more challenging than genotyping. As a result, lifetime environmental exposures cannot be measured as accurately as genotypes. In retrospective studies, such as the most commonly used case-control studies, information on environmental exposures cannot be obtained from direct measurements. Unless there is a biomarker with a sufficiently long half-life, this information is usually obtained indirectly via questionnaires or occupational records. Most of the time, the result is a dichotomous exposure variable that very crudely summarizes the environmental exposure history as a "yes" or "no." Especially when quantitative information is needed, questionnaires do not provide information that is as accurate as that provided by genotyping for genetic exposures. One solution is use of prospective cohort studies, where measurements can be taken as exposures occur, but the resources needed for cohort studies are considerably larger than those required for case-control studies.

There are different types of statistical interactions

When an environmental factor and a genotype jointly exert an effect different from the sum of their individual effects, it is called a statistical interaction. A statistical interaction does not necessarily mean that the environmental factor and genotype interact biologically. **Biological interaction** refers to two factors that act together in a physical or chemical reaction or in a causal mechanism for disease development. Statistical interactions can also be described as heterogeneity of effects because the genotype has a different effect depending on the absence or presence of the environmental factor, and vice versa. This is technically defined as between-stratum effect-size heterogeneity.

When a statistical interaction between the environment and a genotype is mentioned, the type of interaction should be made clear. Just to emphasize that both genes and the environment are important in disease development, **co-action** has been proposed as a better term than interaction. Co-action refers to the independent joint action of different genes and/or environmental risk factors. The practice of describing any interplay between genes and the environment as interaction is not useful, and the word interaction should be reserved for the specific type of interplay detailed below.

Different types of statistical interactions and how they relate to individual effects of the genotype and environmental factors are shown in Figure 5.3. In the first four examples (**Figure 5.3A–D**), there is no interaction between the genotypes and the environmental exposure. In Figure 5.3A there is neither an association nor an interaction, while in Figure 5.3B there is a genetic association, as the odds ratio increases for genotypes with increasing numbers of the minor allele B with no difference in environmental exposure. The opposite is seen in Figure 5.3C, where there is a purely environmental association that does not differ with genotype. Figure 5.3D illustrates a situation that is frequently confused with an interaction. There is a genetic association, as can be seen from the odds ratios for each genotype, and they are different in the presence of the environmental exposure. However, the contribution of the environment to the risk is the same for each genotype, representing purely a joint effect of the genotype and the environment. **Figures 5.3E and F** are examples of statistical interactions. In Figure 5.3E, the odds ratio for each genotype increases by more than a fixed amount due to changing environmental exposure. Figure 5.3F depicts a different type of interaction, where the presence of the environmental exposure changes the direction of the association from protective to risk.

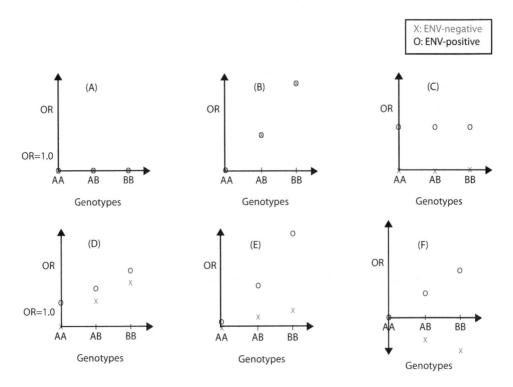

Figure 5.3 Different types of genetic end environmental interplay. (A) Regardless of environmental exposure, there is no association (all ORs are at the unity line); (B) a genetic association with no modification by exposure (no between-stratum effect-size heterogeneity); (C) OR is high in exposed individuals but remains the same for all genotypes (an environmental association with no interaction with genotype); (D) associations with both an environmental factor and genotype with no additional effect over and above joint effects (no interaction); (E) ORs show heterogeneity between the exposed and unexposed strata for each genotype and the association is stronger than the joint effect in the exposed subset (synergistic gene and environment interaction). This is an example of quantitative or removable interaction (log transformation of the ORs would remove the synergistic interaction); (F) the associations are in opposite directions in exposed and unexposed strata (crossover effect). This is an example of qualitative or nonremovable interaction. OR, odds ratio for disease risk; ENV-positive, environmental exposure positive; ENV-negative, environmental exposure negative; genotypes AA, AB, and BB denote wild type, heterozygosity, and variant homozygosity, respectively.

It is important to appreciate the differences between the two types of interaction shown in Figures 5.3E and F. The type of interaction illustrated in Figure 5.3E can be subject to a mathematical manipulation called transformation, which makes it resemble Figure 5.3D. This type of interaction is called a removable interaction (by mathematical transformation of the odds ratio values). The same manipulation cannot remove the interaction shown in Figure 5.3F. This type of interaction is called a nonremovable interaction. Removable and nonremovable interactions are also called quantitative and qualitative interactions, respectively (**Table 5.1**). A removable interaction can be quantified as the magnitude of deviation from a joint effect, but in a nonremovable interaction, the direction of deviation is more important than its magnitude. A nonremovable interaction is of greater interest than a removable one as it indicates effects in opposite directions (also called **crossover effects**). A crossover effect remains hidden in an association study unless specifically

Table 5.1 Alternative terminology, types, and features of statistical interactions

Removable interaction	Nonremovable interaction
Quantitative interaction	Qualitative interaction
Deviations from additive or multiplicative scale (therefore, can be removed by transformation of the odds ratios)	Crossover effects (including sexually antagonistic interaction)
Most commonly considered type	Infrequently considered
Detection may require a very large sample size	Detection does not require a large sample size

explored, because the net effect shows no association in an overall analysis. One example of a crossover effect is a **sexually antagonistic interaction** in which the odds ratios for the genetic associations are greater than 1.0 (denoting risk) in one sex and less than 1.0 (denoting protection) in the other sex. Here, sex acts as an environmental exposure and this is a gene and sex interaction.

The interactions described so far are all synergistic ones in which the odds ratio is greater than the sum (on an additive scale) or product (on a multiplicative scale) of the two odds ratios for genotype and environment effects. An antagonistic interaction denotes a joint effect of the genotype and environmental exposure that is weaker than the sum or product of their individual effects. This should not be mixed with sexually antagonistic interaction, which is a crossover effect as explained above. **Table 5.2** shows how the ORs for synergistic and antagonistic interactions differ from ORs from co-actions of a genotype and environmental factor. Causality of statistical interactions and whether they correspond to biological interactions can only be confirmed by further functional studies.

There is no need for each factor to have an individual (main) effect for a statistical interaction to occur. In the extreme example of PKU, neither the gene mutation nor phenylalanine intake have an effect on health (no main effect), but together they cause irreversible damage to the brain within the first year of life. In practice, most explorations of interaction are restricted to the variables with main effects, which is not a complete exploration.

The environment includes age and sex

Environmental factors include age and sex, which are the most common effect modifiers as well as being common confounders. Examination of each sex and different age groups separately for an association is a good idea in case there are sex- and/or age-specific associations leading to interactions between the genotype and sex or age. Statistical adjustment for age and sex can rule out confounding but will not uncover associations that occur only in one sex or in one age group. It is plausible that the association is in different directions (protective versus risk) in each sex and that the overall effect is zero (no main

Table 5.2 Different scales of removable statistical interactions between a genotype and an environmental factor

	Additive (numerical) scale	Multiplicative (logarithmic) scale
No interaction (co-actions)	$OR_{GE} = OR_G + OR_E - 1$	$OR_{GE} = OR_G \times OR_E$
Synergistic interaction	$OR_{GE} > OR_G + OR_E - 1$	$OR_{GE} > OR_G \times OR_E$
Antagonistic interaction	$OR_{GE} < OR_G + OR_E - 1$	$OR_{GE} < OR_G \times OR_E$

OR_{GE} is the OR for gene and environment interaction, OR_G is the genetic main effect, and OR_E is the environmental factor main effect.

effect but a crossover effect). In this case, only a statistical interaction analysis will uncover the association. For example, recombination rate is associated with the same SNPs (allele C at rs3796619 and allele T at rs1670533) but in opposite directions in males and females. Sexually antagonistic associations like this one may never attain statistical significance in overall analysis. Another possibility is that the association is only present in one sex as the genetically susceptible subtype, and the other sex shows no association. In this case, the magnitude of the strength of the association in the susceptible sex would be the main determinant of the statistical significance in the overall analysis. Such useful observations may be missed if separate analysis for each sex is not performed.

Why don't we see many interactions?

Explorations of statistical interactions seem to be avoided for two reasons. First, the strictly statistical approach dictates avoidance of **multiple comparisons**, which results in a cutting down of the number of analyses. While the concern for spurious associations resulting from an excessive number of analyses is real, there are ways to control false positives. The strictly epidemiologic approach is to do all necessary analyses, repeating all positive findings to rule out the role played by chance, and then to proceed to functional replication studies. Second, the belief that the analysis requires an unrealistically large sample size to uncover an interaction is discouraging. There is some truth in this, but it mainly applies to quantitative interactions where the sample size needs to be four times the sample size used to detect the main effects of each exposure.

Qualitative interactions or crossover effects (including **sexually antagonistic associations** where the association is in the risk direction in one sex and in the protective direction in the other) can be detected in studies designed to detect main effects. Crossover effects can exist even in the absence of main effects; that is, there is no association of the genetic marker or sex with the disease in the overall analysis, but there are opposite associations in each sex. Such effects are more important to recognize than other types of interactions when determining public intervention decisions, as a subset of the population may actually be harmed from a possible intervention if the crossover effect is not recognized. The interacting factor does not have to be sex; any other exposure may interact with a genetic marker to create a crossover effect.

Gene and environment interactions may be exploited to detect genetic associations

If a genotype is a modifier of the effect of an environmental exposure, consideration of the interaction generally increases the likelihood of detecting genetic associations with otherwise small main effects. While all study designs used to detect genetic associations have the potential to detect gene and environment interactions, the case-only design is the most useful provided that the independence of genotype and exposure is assured (**Box 5.2**). For example, if a genotype by smoking interaction is going to be explored in a case-only study, genetic variants that increase tendency to nicotine addiction will not show independence in smokers and nonsmokers in the healthy population from which the control group would be selected. In this case, interactions with those variants cannot be assessed in a case-only study. If genetic and environmental factors are not independent in controls, the case-only analysis will have an inflated type I error rate (false positives). In most scenarios, the statistical power needed to detect an interaction is greater for this design compared with case-control, family triad, or cohort studies. Power is maximized when individual main effects are weak (OR < 1.2) and the interaction is at least moderately strong (OR > 1.5). When the

---

Box 5.2 A case-only study to explore the sex effect (interaction of genotypes with sex) in a genetic association in cancer

Cases with cancer are genotyped.

Rather than cases and controls, male and female cancer cases are compared for genotype frequencies.

The estimated odds ratio is for the interaction of the genotype with sex and reflects the risk change in males compared with females.

The independence of genotype with sex has to be assured in a healthy population sample if it has not already been checked. The frequency of the risk genotype should not differ between healthy males and females.

The genetic main effect (the OR) for the association in males and females cannot be estimated in a case-only study.

---

main effect is weak and the interaction is stronger, the sample size required to detect the interaction is smaller than the sample size required to detect the main effect.

5.3 The Environment and Gene Activity

A gene may have a different activity due to changes in its expression or structure caused by variation in its DNA sequence. This is the basis of genetic associations with phenotypes. Epigenetic variation modifies gene expression levels without any change in DNA sequence. Epigenetics can be seen as the study of the environmental regulation of gene expression and is becoming increasingly important in genetic epidemiology. Lifestyle behaviors, social cues, and dietary, occupational, and pharmacologic exposures all have an impact on the epigenetic status of our genomes. Epidemiologic studies on epigenetic changes do not measure environmental exposure but examine the end result of such exposures.

There are three main epigenetic phenomena (**Figure 5.4**). DNA methylation is a powerful mechanism, especially when it occurs in the promoter regions of genes. Biochemical changes in histone proteins that surround the genes and regulate the access of transcription factors to gene promoters are also epigenetic changes. Another epigenetic mechanism is the control of post-transcriptional processes by microRNAs. The environment can influence all three epigenetic mechanisms. Epigenetic changes make genes with the same DNA sequence behave differently. For example, people who were conceived at the time of the Dutch famine at the end of World War II show different expression levels of the gene *INSIGF* compared to siblings conceived at a different time. The DNA sequences of the two groups are the same, but the group conceived at a time of famine show lower methylation levels at *INSIGF*. The script of the DNA sequence is interpreted differently in different situations.

Overall, epigenetics explains the following observations:

- Despite a very high percentage of DNA sequence identity, humans and chimpanzees are very different species.

- Despite identical DNA sequence, monozygotic twins may have differential susceptibility to complex diseases.

- All cells in our body have the same DNA sequence, but they act in unique ways and form different tissues and organs.

- Cells in our body behave differently depending on our sex and age.

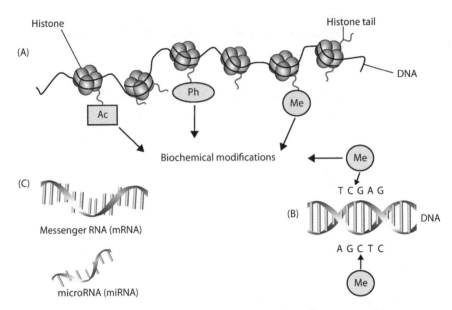

Figure 5.4 Environmentally induced epigenetic changes that regulate gene activity can occur at three levels. (A) Histone proteins may be biochemically modified at their tails; these modifications in turn influence the transcription levels of genes adjacent to them. Acetylation (Ac), phosphorylation (Ph), and methylation (Me) are some of the possible modifications. (B) The cytosine base in nucleotides can be methylated (Me). This modification usually represses the transcription of the gene nearby. (C) After transcription, translation of messenger RNA into a protein product can be repressed by binding of microRNA (miRNA) to mRNA. Environmental exposures modify the level of biochemical modifications at histones or DNA, or the level of miRNAs, resulting in changes in gene expression levels.

Since epigenetic variation contributes to inter-individual variation in gene expression and to differences in complex disease susceptibility, a new subdiscipline in genetic epidemiology has emerged to examine the relationship between epigenetic variation and disease risk: epigenetic epidemiology. Epigenetic studies hold great promise but they are not easy to carry out. The examination of DNA sequence variants is easier because the same variation exists in each nucleated somatic cell. Epigenetic changes, however, are cell-type-specific and may only occur in a unique cell type involved in disease pathogenesis (such as brain cells in psychiatric diseases, or lung cells in smokers). The use of peripheral blood cells as proxies of those changes is being tested but is unlikely to provide a common solution to the necessity of using specific cell types in epigenetic studies. Even within peripheral blood cells there are different cell types, each one with a different epigenetic profile and in a different proportion in different samples. Methods are being developed to adjust for such variation in studies using whole blood samples. Studies on peripheral blood cells have already provided interesting data on epigenetic changes caused by various exposures, and microarrays for **epigenomics** analysis have been developed. Much progress in this field is expected in the near future.

5.4 Gene and Environment Interactions, Personalized Prevention, and Treatment

Gene and environment interactions may have a **public health impact**. The impact of a confirmed gene–environment interaction depends on its strength as measured by the effect size, the frequencies of the environmental exposure, and the interacting genotype. Ideally, the subpopulation at highest risk, those who possess the genotype and are exposed to the environmental factor, should be targeted for public health interventions (such as smokers possessing risk genotypes for smoking-related cancers). However, in some situations, the overall benefit of small changes at the whole population level may be greater than large changes in the high-risk subset. Thus, despite the presence of a known genotype that modifies the detrimental effect of an exposure, a public health intervention may be applied to the whole population without genotyping people. A hypothetical example would be screening the population for *HFE* C282Y mutation positivity and targeting those positive for the mutation for iron depletion, as they are at high risk for iron overload. However, it is more likely that any such intervention would be applied to the whole population rather than the subset at highest risk. There is currently no example of a genotype-based public health intervention, but **public health genomics** is a new and growing discipline to explore such possibilities.

Gene–environment interactions can be useful in a clinical setting

Gene and environment interactions are useful in a clinical setting at the individual level when the exposure is well defined. The best-defined environmental exposures are medicines taken voluntarily. For example, polymorphisms in the cytochrome P450 gene *CYP2C9* and the vitamin K epoxide reductase complex subunit 1 gene *VKORC1* influence the clearance rate of the oral anticoagulant warfarin. In clinical practice, individuals are genotyped in order to adjust drug dosage. Likewise, hypersensitivity to certain drugs is mediated by genetic polymorphisms; for example, hypersensitivity to the anti-HIV drug abacavir is mediated by *HLA-B*57*. As a US Food and Drug Administration approved genetic test, *HLA-B*57* typing is now routinely used to avoid gene and drug interactions. What is expected to become more and more common is the usage of individual genotypes for drug treatment. *HLA-B*57* and abacavir usage is one example for avoiding harm, but individual genotypes may also soon be used to select the most appropriate treatments. In some diseases, such as rheumatoid arthritis, treatment with anti-TNF antibodies can be prescribed to selected individuals based on information available on the genotypes associated with response and nonresponse to the particular treatment.

The *HFE* C282Y mutation increases intestinal iron absorption and causes hereditary hemochromatosis, an iron overload disorder. The genetic effect is strongly modified by environmental factors including dietary or supplementary iron intake, and other factors that modify iron absorption, including alcohol intake. The mutation has low penetrance and only a fraction of individuals who have two copies of the mutation develop hereditary hemochromatosis. Most of those who develop the disease are alcohol users. Up to 15% of people in Western European populations may be positive for the mutation. Given the ongoing iron fortification of food in many countries, and widespread use of iron supplements, screening for the *HFE* mutation has been suggested for individual prevention programs. This proposition does not currently meet with approval for routine clinical use, but provides a good example of a gene and environment interaction that may be used in personalized prevention in the future.

Key Points

- Most diseases have an environmental component.

- Measuring environmental exposure is important to gain maximum benefit from genetic association studies, but it is not an easy task.

- Most environmental exposure measurements are snapshots as proxies for continuous and variable exposure. Cohort studies with repeated measurements provide more realistic information on measured environmental exposures.

- It is important to identify gene and environment interactions, especially when the gene and environmental factors involved do not show individual associations. Unraveling interactions improves accuracy and precision in the assessment of genetic and environmental influences.

- Genetic markers of complex disease susceptibility may be involved in disease development as effect modifiers of environmental factors as well as by direct causation.

- The underused case-only study design provides high statistical power to detect gene–environment interactions, especially when there is no genetic main effect. It has great potential to uncover hidden interactions that require large sample sizes in case-control studies to detect.

- Statistical interactions can be quantitative or qualitative. Quantitative interactions require very large sample sizes to be detected. Qualitative interactions are recognized less, but can be detected in studies of modest size and have greater public health importance.

- Since interactions are possible without main effects, interaction analysis should be performed on variables without main effects.

- The environmental influence on the genome generates epigenetic variation through DNA or histone modifications, which is as equally important as DNA sequence variation in genetic associations.

- Epigenetic association studies are gaining popularity and they examine the end result of overall environmental impact without measuring individual exposures.

- To be able to use genetic association results for potential public health interventions, identification of strong gene and environment interactions is necessary.

Further Reading

Environmental and genetic effects on disease occurrence

Bomprezzi R, Kovanen PE & Martin R (2003) New approaches to investigating heterogeneity in complex traits. *J Med Genet* 40, 553–559 (doi: 10.1136/jmg.40.8.553).

Clayton D & McKeigue PM (2001) Epidemiological methods for studying genes and environmental factors in complex diseases. *Lancet* 358, 1356–1360 (doi: 10.1016/S0140-6736(01)06418-2).

Hemminki K, Lorenzo Bermejo J & Försti A (2006) The balance between heritable and environmental aetiology of human disease. *Nat Rev Genet* 7, 958–965 (doi: 10.1038/nrg2009).

Lander ES & Schork NJ (1994) Genetic dissection of complex traits. *Science* 265, 2037–2048 (doi: 10.1126/science.8091226).

Le Marchand L (2005) The predominance of the environment over genes in cancer causation: implications for genetic epidemiology. *Cancer Epidemiol Biomarkers Prev* 14, 1037–1039 (doi: 10.1158/1055-9965.EPI-04-0816).

Epigenetic epidemiology

Petronis A (2010) Epigenetics as a unifying principle in the aetiology of complex traits and diseases. *Nature* 465, 721–727 (doi: 10.1038/nature09230).

Relton CL & Davey Smith G (2010) Epigenetic epidemiology of common complex disease: prospects for prediction, prevention, and treatment. *PLoS Med* 7, e1000356 (doi: 10.1371/journal.pmed.1000356). (*This review paper provides a background on the epigenome, the genetic and environmental basis of epigenetic changes, how these changes are measured, and whether Mendelian randomization principles can be applied to the study of epigenetic causes of complex diseases.*)

Gene and environment interactions

Dempfle A, Scherag A, Hein R et al. (2008) Gene-environment interactions for complex traits: definitions, methodological requirements and challenges. *Eur J Hum Genet* 16, 1164–1172 (doi: 10.1038/ejhg.2008.106). (*Clearly explains what is meant by statistical and biological interaction, and distinguishes this from confounding.*)

Kraft P, Yen YC, Stram DO et al. (2007) Exploiting gene-environment interaction to detect genetic associations. *Hum Hered* 63, 111–119 (doi: 10.1159/000099183). (*A quantitative example showing how genetic or environmental associations may remain masked until interactions are explored, and which methods are more efficient in which conditions.*)

North KE & Martin LJ (2008) The importance of gene–environment interaction: implications for social scientists. *Socio Meth Res* 37, 164–200 (doi: 10.1177/0049124108323538). (*Despite its title, this paper is extremely useful for any scientist involved in genetic association studies, and especially when the environmental contribution is considered. Using very suitable language, North & Martin explain the basis of genetic association studies, gene and environment interactions, study designs, and technical and analytical aspects of such studies. They also provide practical examples, using obesity as a complex trait.*)

Ottman R (1996) Gene–environment interaction: definitions and study designs. *Prev Med* 25, 764–770 (doi: 10.1006/pmed.1996.0117). (*One of the most informative reviews on gene and environment interactions, written by an eminent epidemiologist. It covers definitions of interaction, measurement of interactions in cohort and case-control studies, and their value in disease prevention.*)

Thomas D (2010) Gene-environment-wide association studies: emerging approaches. *Nat Rev Genet* 11, 259–272 (doi: 10.1038/nrg2764). (*A tutorial on the available epidemiologic designs and statistical approaches for studying specific gene and environment interactions and choosing the most appropriate methods. It clarifies a great deal of information, including different types of interactions, with a lot of real examples. It also covers the study designs for gene-environment interactions.*)

Examination of environmental effects

Morgenstern H & Thomas D (1993) Principles of study design in environmental epidemiology. *Environ Health Perspect* 101(Suppl 4), 23–38 (PMCID: PMC1519688).

Elementary Statistical Concepts

6

Statistical concepts are critical in determining the success of a genetic association study and must be considered before the study starts. Mistakes occurring throughout the various stages of a study cannot always be rectified by statistical tools in the data analysis phase and will result in an invalid study. Once an association is observed, making the distinction between genuine and spurious results is challenging.

The ultimate aim of an association study is to find a specific causal variant directly involved in disease development, not a proxy for it. Statisticians and statistical concepts are invaluable at this phase of the study, and successful research teams are characterized by everyone being familiar with statistical concepts. **Table 6.1** shows a summary of the multiple roles of statistics in genetic association studies. Of these, validity and utility assessment will be the subject of a later chapter. This chapter aims to introduce the basic statistical concepts needed to conduct a successful genetic association study and the Further Reading section provides excellent resources for additional study. Discussions apply to case-control studies, the most common genetic association study design.

Table 6.1 Statistical considerations in a genetic association study

Phase of the study	Statistical issue	Considerations
Planning	Statistical power	How many subjects should be included Ratio of cases and controls Which variants should be included based on their frequencies and functionality Study design considerations such as collection of data on potential confounders for later statistical adjustments and stratification Instrumental variable choice and assessment for Mendelian randomization studies
Post-genotyping, pre-analysis	Quality control	Identification of cryptic relatedness among subjects to exclude related individuals Systematic error due to batch, study center, or cohort effects Genotyping control tests including missingness assessment and Hardy–Weinberg equilibrium
Analysis	Association tests	Single or multiple variant analysis, haplotype analysis Assessment of confounding and effect modification (interaction) Assessment of population stratification Adjustment for multiple testing
Post-analysis	Validity and utility assessment	Causality assessment Clinical validity analysis Genetic risk profiling Potential biomarker development process

6.1 Statistical Power and Statistical Significance

In a genetic association study, a difference in risk marker frequencies between two comparison groups is sought. **Statistical power** estimations ensure that if a difference exists, the statistical tests will reach the predetermined statistical significance level. Otherwise, the difference will be observed but it will not be statistically significant. In statistical language, the power of a statistical test is the probability that it will yield a statistically significant result given that the null hypothesis is false—that is, when there is indeed an association.

Statistical power is expressed numerically and indicates the level of confidence in a study: statistical power of 1.0 (or 100%) means that the study will always detect a true effect if there is one, at the desired statistical significance level. In two different studies, the same degree of difference for the same genetic variant may yield different statistical significance levels due to differences in statistical power: it may remain nonsignificant in one study (insufficient power) and may reach statistical significance in the other (sufficient power).

Calculating statistical power is not a straightforward operation as it depends on several variables, but several software tools are freely available for this purpose. One is Purcell and Sham's online Genetic Power Calculator (GPC), which comes with extensive explanatory notes. Another is the Power for Genetic Association Analyses (PGA) tool, which was developed at the National Cancer Institute (details are given in the URL list at the end of the chapter). These tools are used specifically for calculating statistical power in genetic association studies. General statistical power calculations can be used only as a rough guide because they lack the capability to assess the additional parameters required to calculate the power of genetic studies. The GPC is a commonly used calculator for statistical power and considers all the determinants discussed in this chapter.

Statistical power depends on sample size

In any association study, sample size is the major determinant of statistical power. In practice, statistical power calculations are performed for two scenarios:

1. For a set power level (usually 0.8), what should be the sample size?
2. For a fixed sample size, what would be the power for a certain association?

Figure 6.1 shows how the same effect size is measured with different degrees of statistical significance because of differences in sample size. The confidence intervals (CIs) shown in Figure 6.1 correspond to the plausible range of values the odds ratio (OR) can take. More precisely, the unknown true value of the odds ratio is within the CI at the probability level shown—a 95% CI contains the true value of the OR with 95% probability. If the same study is repeated 100 times, in 95 of the repeats, the OR will fall within the 95% CI. As Figure 6.1 suggests, the size of the CI correlates with statistical significance and gets narrower as statistical significance increases.

Statistical power also depends on effect size and exposure frequency

Statistical power is also affected by the frequency of exposure (such as smoking in environmental epidemiology) and the effect size itself. The higher the frequency of exposure, the higher the statistical power will be. Similarly, the greater the effect size, the greater the statistical power will be. In any association study, the effect size and frequency of exposure are often either unknown beforehand or may have a range of values. They must therefore be estimated using good judgment, or calculations must be made using an appropriate

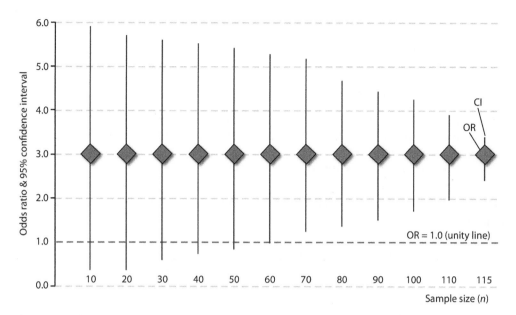

Figure 6.1 Changes in statistical power as demonstrated by the confidence interval (CI) of the effect size (OR) as the sample size increases. In all scenarios, the OR is 3.0, but until the sample size (n) exceeds 60, the lower limit of the confidence interval crosses the unity line (OR = 1.0), denoting no statistically significant difference in the risk. As the sample size increases, the CI gets narrower because of increasing statistical power. Thus, the same association with OR = 3.0 will only be statistically significant if the sample size is 60 or larger, when the lower limit of the CI no longer crosses the unity line. This is what is meant by detecting an association with statistical significance.

range of values. If a complex disease is being studied, the effect size will usually be within the range of OR = 1.2 to 1.5 for individual SNPs, and the range used for power calculations should span realistic values (for example 1.1 to 3.0) rather than unrealistic ones (for example 2.0 to 4.0).

Effect size is measured as an OR in case-control studies (the most common genetic association study design), but different study designs use different parameters: cohort studies measure effect size as relative risk (RR) and longitudinal survival studies give a **hazard ratio** (HR). In clinical trials, effect size can be measured as relative risk, absolute risk change, or number needed to treat (NNT). If the outcome of the study is a quantitative phenotype, such as changes in gene expression levels, one measure of effect size is the proportion of the total phenotypic variance that is caused by the expression quantitative trait locus (eQTL). This value is usually less than 5%.

In situations where a difference between two means is of interest, which would be typically assessed by the t-test, there are a number of effect-size measures. The simplest is the ratio of the change between the means ($\mu_1 - \mu_2$) to the standard deviation (σ), which is called Cohen's d. To obtain this value, the difference between the two means is divided by either of the two standard deviations. If the two standard deviation values are not similar, using one or the other may result in rather different effect sizes, and so the mean of the two values, σ', may be used as the pooled standard deviation:

$$\sigma' = [(\sigma_1^2 + \sigma_2^2)/2]^{\frac{1}{2}}$$

The conventional values for d, as defined by Cohen are as follows:

small $d = 0.2$

medium $d = 0.5$

large $d = 0.8$

Thus, a medium effect is worth half a standard deviation difference between the two means. Similar to unknown ORs before a case-control study, the exact values of Cohen's d will be unknown before a study, and for statistical power calculations these values may be used to obtain a range of statistical power estimates.

The statistical power of genetic association studies depends on the genetic risk model

Frequency of exposure equates to frequency of **risk genotype** in genetic association studies: the higher the risk genotype frequency, the higher the statistical power. It is important to note that it is not the allele frequency but the risk genotype frequency that is used in statistical power calculations. Allele frequencies are converted to risk genotype frequencies using a genetic risk model (see Chapter 4) and may be a single genotype or two genotypes pooled together. For example, in the dominant model, the variant allele (B) is present in both heterozygote (AB) and homozygote (BB) genotypes and the risk genotype frequency is the sum of the heterozygote and homozygote genotype frequencies. In the recessive model, the variant allele (B) is present only in the homozygote (BB) and the risk genotype frequency is the square of the allele B frequency. Thus, the risk genotype frequency depends on the genetic risk model; using different models for the same SNP will generate a different risk genotype frequency.

As the genetic risk model equates to the underlying mode of inheritance, it is usually unknown beforehand and estimations of statistical power are carried out for all possible models. In general, the additive model has the maximum power to detect an association whatever the underlying risk model is. If the actual model is recessive, the association may be missed in a study powered to detect associations fitting to an additive model. Therefore, the recessive model may not be detected by exclusive use of additive model analysis (as is usually done), especially when the risk genotype frequency is not high.

Disease prevalence and linkage disequilibrium are also important for statistical power in genetic association studies

Genetic association studies examine multiple SNPs in one study. The SNPs included will have varying frequencies and statistical power calculations should therefore be made for a range of frequencies. Statistical power is greatest when the allele frequencies are within the intermediate range: rare (less than 5%) or common (close to 50%) alleles yield lower statistical power for resulting risk genotypes. However, with a common (prevalent) disease, high-risk variants are assumed to be lacking in the control group selected to be disease-free, while cases are expected to be enriched for high-risk variants. This large difference in the frequency of alleles between case and control groups increases statistical power. For rare diseases, no large divergence is possible to help with statistical power.

Linkage disequilibrium between the marker being examined and the unknown causal variant must also be considered; however, this must be estimated as it is unknown.

If there is strong evidence that the causal variant is within a small chromosomal region and a high number of SNPs from that region will be screened, it may be safe to assume there will be absolute LD ($D' = 1.0$; $r^2 = 1.0$). In this case, the allele frequencies of the two variants will also be identical. If no prior information is present or, as in the case of a genome-wide association study (GWAS), when there are so many markers, it is safer to calculate statistical power for a range of potential LD values. It is safest to use $r^2 = 0.80$ with some confidence rather than to assume $r^2 = 1.0$ for all variants. As r^2 gets smaller, the variant being examined will be less representative of the causal variant, and the statistical power will be lower. Representativeness is an issue in a GWAS when common variants that are presumed to be proxies for usually rare causal variants are used. The frequency difference between the examined (common) variant and the unknown (rare) variant means there is a low r^2 value for LD between them.

6.2 Study Design and Errors

Sampling variability affects the statistical power of a study

If any association study is repeated many times in the same population, the result will not always be the same due to sampling variability. Random variation occurs and chance may generate spurious results. If there is a true effect (the null hypothesis is wrong and the alternative hypothesis is true), sampling variability will prevent the detection of this true effect 100% of the time, but ideally the true effect should be detected most of the time when the same experiment is repeated.

It may sound sensible that all studies should aim for full power so that they will detect any existing association every time they look for it, but this can only be achieved if sampling variability is eliminated by including almost all of the population from which the study sample would have derived. This is not practical and it is generally agreed that statistical power of 0.8 is a good compromise and is the optimal power for most studies. **Figure 6.2** shows the relative increase in sample size needed to obtain statistical power of more than 0.8. In a study with power 0.8, if a true effect exists, it will be detected 80% of the time and will be missed 20% of the time (false negativity). These values are agreed upon to be acceptable generally.

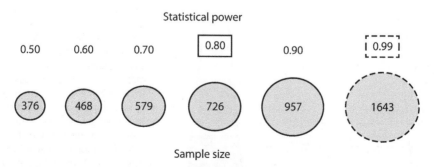

Figure 6.2 The sample size required to yield a range of statistical power when the risk genotype frequency and effect size are kept the same. The example is based on a risk genotype frequency of 0.10 in controls and 0.15 in cases, with equal numbers of cases and controls, and a statistical significance threshold of 0.05.

Association studies are a trade-off between sensitivity and specificity: type I versus type II errors

Failing to detect a true effect with statistical significance is a false-negative result—statistically speaking, a **type II error**. Type II error probability is denoted by the Greek letter β and estimated as (1 − statistical power). Because of the relationship between statistical power and type II errors, if the statistical power is less than 0.5, the false-negative result rate will be greater than 50%. The **statistical significance threshold**, known as the α value or **type I error** (false-positivity rate), represents the probability of obtaining a false-positive result. By convention, the accepted statistical significance threshold (false-positive rate) is an α value of $P = 0.05$, meaning that 5% of the time, a nonexistent effect will be found to be statistically significant. The **P value** is a statistical function that estimates the reliability of data collected to test a hypothesis—the probability of obtaining a false-positive result due to random variability.

The accepted level of statistical power in any association study is 0.80 (β = 0.20). Setting the α value at $P = 0.05$ and statistical power at 0.80 is a trade-off between **specificity** and **sensitivity**. **Figure 6.3** illustrates the relationship between the two types of statistical errors. If the statistical significance threshold (α) is lowered by setting it at, for example, $P = 0.10$, sensitivity increases (more associations will be statistically significant and there will be fewer false negatives) but specificity decreases (there will be more false positives). If the significance threshold is set more strictly, for example at $P = 0.01$, the study will have less power to detect associations reaching this significance value. The result will be fewer associations with statistical significance and more associations will be missed (more false negatives), but there will be fewer false-positive results.

When power is too high because of a very large sample size, even trivial differences may reach statistical significance. In a study with a small sample size, only large differences

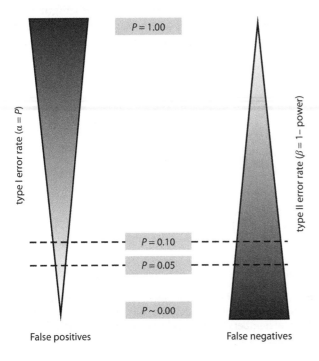

Figure 6.3 Trade-off between type I and type II error rates. As the statistical significance threshold is lowered to minimize false-positive results, most detected associations are real. However, more of the existing associations—such as those with rare variants—may be missed due to the P value for their association not reaching the more stringent threshold (a higher false-negative rate). On the other hand, if the α value is set for lower stringency (say, $P = 0.10$), then there may be no false negatives but more associations reaching statistical significance at this level will be false positives.

between cases and controls will reach statistical significance, but in a study with a large sample size, even very small differences may result in a statistically significant finding. For example, in a study with a sample size of 35,000 cases and 35,000 controls, risk genotype frequencies of 0.10 in controls and 0.11 in cases would yield a statistically significant association with an α value of <0.001. Such high-power studies may find statistically significant results but the **biological significance** of the results may be doubtful.

An α value of 0.05 is the default in most association studies where one, or a few, comparisons are made. However, a GWAS involves comparisons of millions of frequencies. A high number of comparisons increases the risk of false-positive findings and so some statistical safeguard against undesirable false positivity is needed. The simplest adjustment involves lowering the statistical threshold for significance. In a GWAS, it is assumed that one million independent comparisons are made even though the actual number of comparisons may be higher (some comparisons are for SNPs that are correlated to others and therefore not independent). If one million comparisons are made, the conventional threshold is corrected for this number by a simple division ($\alpha = 0.05/1,000,000 = 5 \times 10^{-8}$). Everything else being equal, changing the α level from 0.05 to 5×10^{-8} requires a substantially larger sample size to achieve the same degree of statistical power. As a rule of thumb, the sample size should be increased roughly in proportion to the log change of the α value to retain statistical power as α decreases. For example, the sample size should be doubled as the α value changes from 0.05 to 0.005, which corresponds to a ten-times or one-log change in α value.

There is no measure that will prevent all type I and type II errors; they can only be minimized. Study design must also balance the two types of error, as any modification that reduces one type of error may increase the probability of the other type, given that statistical power remains the same (see Figure 6.3). By using full statistical power (1.0), the type II error rate may be reduced to zero, meaning that there will be no false negatives (all existing associations will be detected with statistical significance), but the false-positive rate will increase.

Traditionally, type I errors (false positives) are deemed to be more harmful and researchers have leaned more heavily toward avoiding them. One of the safeguards against false positives is the statistical correction for the number of tests performed, since multiple comparisons are the main source of false-positive results. This policy certainly avoids false positives but, in turn, it increases the false-negative result rate. Both types of errors can be reduced by increasing statistical power and reducing the statistical significance threshold, but this comes at a large cost in terms of the resources required to use a larger sample size. The example given in **Table 6.2** is based on an association with an OR = 2.0, risk genotype frequency of 0.10 in controls, and equal case and control numbers. The stringent approach requires a large increase in sample size. For such an association to be detectable with 0.8

Table 6.2 The trade-off between type I and II errors and statistical power

	α Value	Statistical power	Sample size
Conventional approach[a]	0.05	0.80 (β = 0.20)	518
Stringent approach[b]	0.01	0.95 (β = 0.05)	2348

[a]Type I error rate (α) is 5% and type II error rate (β) is 20%; [b]type I error rate (α) is lowered to 1% and type II error rate (β) is lowered to 5%.

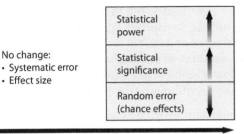

No change:
• Systematic error
• Effect size

Sample size (small to larger)

Figure 6.4 Sample size correlates with different parameters. The increase in sample size is no assurance against systematic errors (bias and confounding), and the effect size of a true association remains the same although its statistical significance increases. A larger sample size provides greater statistical power (therefore, lower type II error or false-negativity rate) and reduces the probability of chance effects (therefore, decreased type I error or false-positivity rate).

statistical power at the conventional statistical significance threshold, the total sample size needed is 518. When both α and β values are reduced for lower false-positive (α) and false-negative (β) rates simultaneously, the required sample size increases more than fourfold.

It has to be remembered that focusing exclusively on statistical threshold reduces the probability of chance findings (random error), but it is no safeguard against spurious findings caused by systematic errors such as bias or confounding (**Figure 6.4**). A biased study can still generate a very small P value that may never be confirmed in following studies.

Insufficient statistical power means undesirable consequences

Statistical power is the probability of detecting statistically significant differences; in studies lacking power, even a true effect will not attain statistical significance. A study may find an association with an OR of 2.4 that looks impressive but with an accompanying P value of 0.09. This may be a sign of insufficient statistical power—the OR is good but not statistically significant. The finding will be considered a chance finding, but it could have been a statistically significant finding if the study had sufficient statistical power. A power calculation based on realistic assumptions (especially the effect size, risk genotype frequency, and LD between the markers to be genotyped and the unknown causal variant) is a strict requirement before the study begins. There is no remedy for low power after the study is completed, but the finding may still contribute to a statistically significant association in a future meta-analysis. Before designing a study, the sample size needed to detect the required effect (for example, an effect size with OR = 1.5) with the desired statistical significance (for example, $P \leq 0.05$) should be worked out. In **Table 6.3**, this point is illustrated by showing how statistical power increases and the P value gets smaller for the same effect size and risk genotype frequency as sample size increases.

Table 6.3 For frequencies of a risk genotype of 20% in cases and 10% in controls, statistical power and statistical significance change with sample size for the same effect size

Effect size (OR = 2.25) and 95% CI	2.25 (0.63 to 9.66)	2.25 (0.93 to 5.77)	2.25 (1.11 to 4.72)	2.25 (1.22 to 4.23)	2.25 (1.31 to 3.92)
Sample size[a]	50	100	150	200	250
Statistical power $(1 - \beta)$[b]	0.20	0.43	0.62	0.76	0.85
P value (α)[c]	0.26	0.07	0.02	0.007	0.003

[a]Number in each study arm for equally sized case and control groups; [b]β is the type II error rate; [c]two-sided P value estimated by Fisher's exact test.

It is common to see low-powered studies yielding good ORs that are not statistically significant and authors stating that if the sample size were doubled, the result would have been statistically significant. If the association was a true association, this may be the case. However, most such findings in small studies are chance findings that will disappear when the sample size increases (and random error decreases). The thought that nonsignificant results will become significant with increasing sample size is therefore not always valid.

A negative finding in an underpowered study is not a true negative finding. For example, in a study with insufficient statistical power (say, 0.50), an OR = 2.4 with an accompanying P value of 0.08 cannot be considered a truly negative finding. It would have been a truly negative finding if the statistical power was >0.80 and the P value was still >0.05. In a study of a disease that is more common in males, there will be fewer females in the sample. Even if the effect size is the same for the association in males and females, the result in females may not reach statistical significance. Such a result should be considered inconclusive rather than negative since the study did not have sufficient statistical power, and the lack of statistical significance in the female group is not unexpected. A negative result is more likely to be true when the study is well powered and the findings do not reach statistical significance. Thus, a good study design at the outset should ensure stronger inferences from both positive and negative results.

Study design can be modified for more powerful studies

There are ways to increase statistical power without increasing sample size. If the aim is to find genetic determinants of height, a study on 10,000 randomly selected people can examine correlations between genotypes and height; alternatively, 1000 extremely short and 1000 extremely tall people can be used. By doing the latter, the difference between the two study groups is inflated to increase the effect size, resulting in a high-power study that will be better able to detect true genetic associations with height. Another method to increase divergence in the frequency of risk markers between case and control groups is to use a subgroup of cases enriched for the genetic determinants of the condition. A classic example is to use young-onset cases in a cancer association study since genetically determined cancer occurs at younger ages. Other ways to enrich the sample with genetically loaded cases include, for example, examining only non-obese subjects with type 2 diabetes, or cases with family history or with aggressive forms of the disease. In all these examples, the case group of the sample has been enriched for cases with higher genetic loads by focusing on hypothesized genetically determined subsets. Using a genetically loaded subset of cases is called **sample enrichment**. It is important to realize that this process may lead to selection bias, and caution is therefore needed. Also, the results will not be applicable to the general population but only to the subgroup examined. The aim of studies using sample enrichment is primarily gene discovery. **Hypernormal controls** believed to be free of genetic determinants for the condition under study can also be used. An example would be using controls beyond the age of onset of a condition in order to minimize the probability that some controls have the genetic determinants for the condition.

The effect size can also be increased to obtain higher power by reducing measurement errors with more accurate phenotyping, or by genotyping a greater number of markers in a specific region of interest to increase the likelihood of getting close to the causal variant. The first instance refers to decreasing heterogeneity of the phenotype. For example, a study on genetic risk factors of asthma may include all types of asthma or just a subtype, such as exercise-induced asthma, since each subtype may have different risk factors.

The second approach aims to increase the unknown LD level between genotyped variants and the unknown causal variant, which would have yielded the greatest effect size. Variants near the causal variant, and in LD with it, will yield smaller effect sizes in proportion to the magnitude of LD with the causal variant.

Increasing the number of control subjects also adds to statistical power, especially when the case numbers cannot be increased due to rareness of the disease. The optimal ratio of case to control numbers is 1:4. Increasing the controls to more than five times the number of cases will result in diminishing returns, as the cost of doing the study increases more than the statistical power.

If applicable, there are also advanced methods to further increase statistical power at the analysis phase of a genetic association study, such as haplotype analysis to account for untyped rare causal variants, pooling of all rare variants in a gene, or pooling of all alleles in genes within the same pathway (pathway analysis). These approaches may provide higher power by increasing the effect size of observed associations but are only applicable at the analysis phase.

6.3 Data Analysis

Statistical analysis is used for quality control before association analysis

Once the raw data are generated, the first task is to verify the **internal validity** of the data. Are the results what they are supposed to be, or have there been some distortions from true results? This begins with quality control checks. It may appear that genotyping quality is a major issue, but the sample and the generated results are also subject to quality control steps.

The most common statistical approach to genotyping quality control is Hardy–Weinberg equilibrium (HWE) testing (Chapter 3). If HWE is found to be violated in the randomly selected healthy control group, the first step is to exclude any biological reason. Plausible biological causes of HWE violation include inbreeding (usually in a small, isolated population), population structure, and selection. If these causes are not considered possible, **genotyping errors** must be explored (the topic of the next chapter).

The sample should consist of only unrelated individuals. The **cryptic relatedness** of subjects within the sample is tested before moving on to the analysis of association. The genome-wide genotype data generated for disease association testing is used for this purpose. Since related individuals may distort the results, those exceeding a certain threshold of genetic relatedness in pairwise comparisons are excluded from the study.

The log **quantile-quantile (QQ) plot** is a useful statistical tool for interpreting the results before specific association analysis. A QQ plot is a graphical technique commonly used in mathematics for determining if two sets of numbers come from populations with a common distribution. This graph visualizes the observed results against the expected ones to check whether there are any systematic deviations between them.

The tests briefly introduced here and outlined in Table 6.1 are discussed in further detail in later chapters. These pre-association tests are indicators of the quality of the sample, genotyping, and results, and only if these tests yield satisfactory results can association analysis can begin.

Logistic regression is the most common statistical test for the analysis of case-control studies

Association analysis begins with the selection of the appropriate statistical test. For the case-control design, this is usually **logistic regression**, a regression method designed for studies that have a binary outcome (such as case or control in a case-control study). Logistic regression is performed using raw data and requires statistical software. There are many widely available statistical analysis packages—for example, SPSS, Stata, SAS, and R—that can be used. Using logistic regression allows the OR to be adjusted for **potential confounders** and enables joint analysis of multiple risk markers. An **interaction** analysis can easily be performed by logistic regression. This is important because one of the most common statistical errors in genetic association study reporting is to not provide a statistical assessment of an interaction for claims of differences in association strength between strata (like sex- or age-specificity of an association).

Fisher's exact test and the Chi-squared test can be used in association analysis

If no statistical adjustment of the OR is necessary, simple tests using the counts of genotypes in cases and controls can be used to assess associations. Most small-scale and purely genetic studies can be analyzed by constructing 2×2 tables for genotype counts (wild type versus risk genotype) in cases and controls (**Box 6.1**).

Box 6.1 A 2×2 analysis of results in a simple case-control study

In a hypothetical genetic association study, 200 cases with type 2 diabetes and 200 healthy controls were genotyped at rs123456 C > T. Genotype frequencies were as shown in the 2×3 table below:

	CC (wild type)	CT (heterozygote)	TT (homozygote)
Cases	27	79	44
Controls	49	71	30

The table above can be rearranged for dominant model analysis to the following 2×2 table, by collapsing genotype counts of CT and TT into a single category:

	CC (referent)	CT + TT (risk genotypes)
Cases	27	123
Controls	49	101

The analysis of this 2×2 table by Fisher's exact test yields a P value of 0.005 (which can be replicated using an online statistical calculator such as GraphPad QuickCalc). Given that $P < 0.05$, it can be concluded that the difference in genotype frequencies between cases and controls is statistically significant and there is an association. The OR is 2.2 and its 95% confidence interval is 1.3 to 3.8. This OR is unadjusted. If age or sex is a potential confounder, it needs to be adjusted. For that purpose, it is possible to use multiple 2×2 tables and a special method to calculate the adjusted OR; this may become cumbersome for more than one confounder. More conveniently, logistic regression provides an adjusted OR for one or even more potential confounders.

Such a table is called a contingency table and can be analyzed by standard statistical tests such as a Fisher's exact test or a Chi-squared test. If any expected count is less than five, it is not appropriate to use the Chi-squared test. Fisher's exact test is suitable for small and moderate sample sizes and can be used to analyze almost all comparisons in a study without additional corrections. However, calculation of P values by Fisher's exact test can be cumbersome because the formula uses factorials of genotype counts, which becomes rather hard to do without computers if the sample size is large. This is why this test has acquired a reputation for being suitable only for small studies. However, given the powers of even low-end personal computers, Fisher's exact test can now be used in all analyses and it is better to use the exact test for all comparisons in small to moderately sized studies. For larger sample sizes, Fisher's exact test and a Chi-squared test yield almost identical results and no correction is needed for the Chi-squared test.

The P value has a function but does not prove anything

Most research results are assessed by the statistical significance level, α, given by the P value. The accepted standard of $P < 0.05$ simply means that the difference observed between the two groups would only occur less than 5 times out of 100 as a result of random error (chance). A P value of 0.05 may be perfectly acceptable in some disciplines, but in biomedical sciences $P = 0.05$ is the largest acceptable value. One very common mistake is reporting a P value as 0.000. Some software may report very small P values as 0.000, but it only means $P < 0.001$ and should not be reported as 0.000. It is also not uncommon to see P values reported as 0.04909 or 0.01218. This is **spurious accuracy** and is not good practice. Retaining too many decimal points does not make the value any more meaningful. The P values in these examples are best reported as 0.05 and 0.01.

Putting too much emphasis on the P value may cloud the consideration of biological importance over statistical importance. A statistically significant result may not be biologically or clinically significant. In any case, the P value cannot be used to declare any statistical correlation as evidence for causality, although the greater the statistical significance, the more likely that the correlation is causal. However small the P value is, what matters is the absolute change in the risk. Because of their large sample sizes and resulting high statistical power, GWAS results from collaborative consortia are extremely statistically significant, with P values reaching values as small as $<10^{-500}$. The OR for some of these results is as small as 1.04, denoting a very small change in risk. Such results may still have biological significance given that a SNP is only one of many interacting factors, but only large ORs will prompt any action toward drug or vaccine development, public health intervention, or implementation of predictive tests.

Epidemiologists emphasize the effect size more than the P value

In epidemiology, the focus is on the effect size, which denotes the strength of an association that correlates with the impact on public health. When presented with a 95% CI, the effect size provides reliable information on how strong the association is, its statistical significance, and its robustness. The more distant the OR from the unity line (for example, >3.0 for risk or <0.35 for protective associations) and the narrower its 95% CI (2.8 to 3.2 rather than 1.1 to 4.9), the more robust and statistically significant the results are. To be valid, however large or small it is, the OR should still reach statistical significance. In fact, if a large OR is not statistically significant, such a result may be due to insufficient statistical power, perhaps due to a small sample size in a study of a rare variant. **Table 6.4** provides an example of how P values change with different genotype frequencies in cases while the OR remains the same.

Table 6.4 Fixed sample size and odds ratios, different risk genotype frequencies in cases, and resulting P values

Odds ratio	1.5	1.5	1.5	1.5	1.5	1.5
Risk genotype frequency (cases)	0.50	0.40	0.30	0.20	0.10	0.01
P value	$<10^{-5}$	2.6×10^{-5}	4.6×10^{-5}	0.0004	0.016	0.63

A large effect size of an association suggests that this finding can be used for screening to find high-risk individuals, or that any intervention based on this association will have a greater impact on public health. For example, smoking increases the risk for lung cancer by more than 10 times, and carcinogenic human papillomavirus (HPV) subtypes increase cervical cancer risk by almost 100 times. Such large effect sizes lead to public health action, for example anti-smoking campaigns and vaccine development for HPV. However, small effect sizes are not always inconsequential.

If an enzyme gene polymorphism shows a statistically highly significant and replicated association but with a small OR due to a subtle change in the enzyme function, this information may be exploited for drug development. By changing the function of the enzyme more strongly than the polymorphism does, a drug may confer a large therapeutic benefit. For example, HMG-CoA reductase encoded by *HMGCR* is a rate-limiting enzyme involved in cholesterol metabolism. Its polymorphisms (*ScrF*I restriction polymorphism, 8302 A > C SNP, and (TTA)$_n$ repeat polymorphism) were examined in early studies and associations were found but were only of modest significance. HMG-CoA reductase was already known to have an effect on cholesterol levels and its inhibition by statins has had a revolutionary impact in the medical care of patients with high cholesterol levels.

Negative results are as valuable as positive ones

Scientific research tests a hypothesis and generates evidence for or against the hypothesis. Obviously not all hypotheses will be true. There is therefore no shame in finding a result that is statistically nonsignificant. A negative result is as important and useful as a positive one, but many researchers are reluctant to report negative findings. No one should avoid reporting statistically nonsignificant findings and accepting a null hypothesis if that is what the results show. Trying to turn negative findings to positive findings by overzealous data dredging and subsequent subgroup analysis is not good practice. Not all positive findings are real and chance must always be discussed as an alternative explanation for any finding. Nothing is proven by a statistically significant result, definitely not causality. Statistics only provides support for or against a hypothesis. It does not prove or disprove anything.

Key Points

- An association study should be sufficiently powered (0.8) to obtain statistically significant results. An underpowered study may show seemingly large differences that will not reach statistical significance.

- The most controllable determinant of statistical power is sample size, which also reduces random error as it gets larger. However, large sample size is no assurance against systematic error (bias and confounding).

- A promising but statistically nonsignificant result may reach statistical significance in a well-powered second study or, if it was due to chance, may remain nonsignificant.

- A nonsignificant result is a truly negative result only when the study is well powered. Inconclusive results may be avoided by designing studies with large enough sample sizes.

- The maximum possible sample size should be used for high-powered studies. If the disease is rare, the control group should be up to four times as large as the case group for increased statistical power.

- Effect size, risk genotype frequency, and linkage disequilibrium also affect statistical power.

- Effect size is unknown before the study, but can be maximized by using sample enrichment methods or hypernormal controls. Reducing disease heterogeneity should also help with increasing effect size.

- The unknown LD value can be maximized by increasing variant density in candidate gene regions.

- Simple association analysis can be performed by the Chi-squared test or Fisher's exact test (to generate unadjusted or crude ORs), but logistic regression is needed for the statistical adjustment of results from a case-control study.

- Statistical analysis of the data is not all about examination of genetic associations. There are initial quality control tests to be done, in particular the QQ plot in a GWAS, to assure sample and genotyping quality before proceeding to the analysis of associations.

- Statistics do not prove anything. A statistical result provides evidence for or against a hypothesis.

- Statistical significance should not be mixed up with biological or clinical significance.

- A truly negative result is as useful as a positive one.

- No statistical evaluation is complete without the assessment of plausible interactions and effect modifications.

URL List

CaTS Power Calculator. Center for Statistical Genetics. University of Michigan. http://www.sph.umich.edu/csg/abecasis/CaTS/reference.html (*For two-stage association studies.*)

Power for Genetic Association Analyses (PGA). Division of Cancer Epidemiology & Genetics. NIH National Cancer Institute. http://dceg.cancer.gov/tools/design/pga

Purcell S & Sham P (2003) Genetic Power Calculator (GPC). http://pngu.mgh.harvard.edu/~purcell/gpc

QuickCalcs. GraphPad. http://www.graphpad.com/quickcalcs/contingency1 (*A statistical calculator for 2 × 2 table analysis.*)

Further Reading

Introduction to statistics

Greenhalgh T (1997) How to read a paper. Statistics for the non-statistician. II: "Significant" relations and their pitfalls. *BMJ* 315, 422–425 (doi: 10.1136/bmj.315.7105.422).

McDonald JH (2009) Handbook of Biological Statistics, 2nd ed. Sparky House Publishing. http://udel.edu/~mcdonald/statintro.html (*Originally prepared as class notes for the Biological Data Analysis course in the University of Delaware, these notes are specifically for biology students with minimal mathematics experience. The focus is on statistical tests and how to choose the best one for a biological problem. A spreadsheet to perform almost every statistical test is provided, with the data already entered.*)

Sinsheimer J (2011) "Statistics 101"—a primer for the genetics of complex human disease. *Cold Spring Harb Protoc* 2011(10), 1190–1199 (doi: 10.1101/pdb.top065870). (*A brilliant introduction to basic statistics relevant to the concepts presented in this chapter.*)

Swinscow TD (1997) Statistics at Square One, 9th ed. BMJ Publishing Group. http://www.bmj.com/about-bmj/resources-readers/publications/statistics-square-one (*A freely accessible online book covering the essentials of basic statistics.*)

Statistical power

Evans DM & Purcell S (2012) Power calculations in genetic studies. *Cold Spring Harb Protoc* 2012(6), 664–674 (doi: 10.1101/pdb.top069559). (*A detailed discussion of statistical power in genetic studies with different designs.*)

Hong EP & Park JW (2012) Sample size and statistical power calculation in genetic association studies. *Genomics Inform* 10, 117–122 (doi: 10.5808/GI.2012.10.2.117). (*A clear and concise guide to statistical power calculation that considers different genetic risk models and various scenarios including case-control ratios.*)

Menashe I, Rosenberg PS & Chen BE (2008) PGA: power calculator for case-control genetic association analyses. *BMC Genet* 9, 36 (doi: 10.1186/1471-2156-9-36).

Statistical analysis

Balding DJ (2006) A tutorial on statistical methods for population association studies. *Nat Rev Genet* 7, 781–791 (doi: 10.1038/nrg1916). (*A more comprehensive and detailed presentation of the ideas discussed in this chapter.*)

Goodman S (2008) A dirty dozen: twelve *P*-value misconceptions. *Semin Hematol* 45, 135–140 (doi: 10.1053/j.seminhematol.2008.04.003). (*A commentary on a dozen of the common misinterpretations of the meaning and usage of the P value in scientific reports that also explains why each one is wrong.*)

Lewis CM (2002) Genetic association studies: design, analysis and interpretation. *Brief Bioinform* 3, 146–153 (doi: 10.1093/bib/3.2.146). (*Another very clear guide on genetic association studies and statistical aspects, especially on genetic models.*)

Genotyping Methods and Errors

7

Genetic association studies use data on the genetic variations that occur between individuals (genotypes), and these data are generated by genotyping. Most common genetic variants are single nucleotide variants (that is, SNPs) and, by genotyping them, the two nucleotides at that position in the genotype are determined. It is generally assumed that SNPs provide sufficient coverage for variants that are not SNPs, including repeat polymorphisms such as **short tandem repeats (STR)** and **variable number tandem repeats (VNTR)**. This chapter will provide background to various genotyping methods and then discuss how errors in genotyping are detected and minimized.

7.1 Genotyping Principles

There are many methods for genotyping, some of which require only simple amplification of the region that contains the SNP, while others include an additional step to identify the nucleotides at the SNP site. The most suitable genotyping method for a particular study depends on the type of SNP and the type of study, and especially on the quality and quantity of DNA available. Different SNPs suit different methods because of the principles of molecular genetics outlined in Chapter 1. **Table 7.1** summarizes the most important issues in determining which genotyping method to use; details of the methods are given later in the chapter.

SNPs have different physicochemical properties

The nucleotides to be differentiated by the genotyping process have different physical and chemical properties. These differences determine how easy it is to distinguish them. The most common nucleotide substitution is C > T (**class I SNP**) and these two nucleotides have the most different properties; therefore, a class I SNP is the easiest to identify by any method. An A > T substitution (**class IV SNP**) is the hardest to distinguish because the chemical difference between the two nucleotides is minimal. Genotyping methods based on differentiating these nucleotides at the amplification phase of the process are therefore very error-prone. For class IV SNPs, it is safer to choose a method that uses an additional identification step following the initial amplification. Other SNP classes, **class II** (C > A) and **class III** (C > G), lie in-between the C > T and A > T substitutions in terms of the difficulty level for differentiation.

7.2 Genotyping and PCR

Most genotyping methods involve an initial polymerase chain reaction (PCR) amplification step, and some aspects of PCR are relevant to successful genotyping. Most important is the design of the **oligonucleotides** used in amplification (**primers**) or in the detection of the nucleotides within an amplified fragment (**probes**). For DNA polymerase to initiate

Table 7.1 Issues relevant to genotyping method selection

Issue	Example	Solution	Method of choice
Quality of DNA	Poor-quality, degraded DNA	Methods that do not require long DNA fragments for genotyping	Short-amplicon melting-curve analysis
Quantity of DNA	Small amount of DNA available	Either whole-genome amplification of the available DNA before genotyping, or use of a method that requires small amounts of DNA	Methods that can genotype many SNPs in one assay (including microarrays)
Differences in physicochemical characteristics of each nucleotide	A > T substitution	Methods that do not exclusively rely on distinguishing the two nucleotides making up the genotype	TaqMan® assay that uses oligonucleotides to recognize not just the alleles but also flanking nucleotides
Composition of the stretch of the DNA to be amplified	Presence of CpG islands Presence of repeated sequences	Assay design modifications	Moving the oligonucleotides used to amplify the DNA fragment beyond the CpG islands/repeated sequences, and using special enzymes for amplification
Paralogs or copy number variation	Genes with duplicated copies (like heat shock protein genes) or genes within regions of copy number variation	Selective amplification of the copy of the gene of interest prior to genotyping	Any method on a selectively amplified gene fragment
Only a few SNPs to genotype and/or a small number of samples to genotype	A small candidate-gene study	A manual method may be used for genotyping as opposed to an automated assay	One of the conventional (non-GWAS) methods
Too many SNPs to genotype and/or a large number of samples to genotype	Genome-wide association study	A microarray that is able to genotype millions of SNPs	A microarray-based genotyping platform

DNA replication, a primer has to anneal to one strand of the DNA. Each PCR assay requires a pair of primers annealing to opposite strands of a DNA duplex, with their 3′ ends facing each other. PCR primers are short (18 to 25 nucleotides), single-stranded, synthetic oligonucleotides that are designed to form specific and stable bonds with their target sequence. While designing primers appears to be a simple issue, in practice there are potential difficulties to overcome and there are many rules that govern the design of primers.

Primer sequence and position are important in genotyping

Primers that are designed to amplify a DNA fragment containing a SNP can anneal anywhere outside the immediate vicinity of the SNP (**Figure 7.1**). Primers for use in genotyping experiments must be checked for sequence identity and specificity to their annealing sites. They should ideally have 100% complementarity to their binding sites and should not bind to any other sequence in the genome. This analysis should also rule out the presence of any SNPs within the binding sites of primers, as they would interfere with the specific binding of the primer to its target sequence. Certain primer–template mismatches or

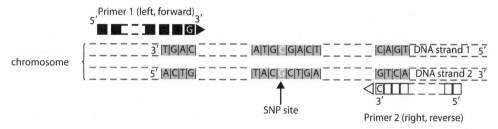

Figure 7.1 Primer location for amplification of a DNA sequence containing a SNP. The primers anneal to each DNA strand in opposite directions. The 3′ terminal nucleotides of the primers must match their target, but mismatches at the 5′ ends are inconsequential. Primers designed to amplify a fragment containing the SNP can anneal anywhere outside the immediate vicinity of the SNP.

mispairings (also called wobbles) may still allow primer extension and DNA replication. Mismatches between primer and template that are most tolerated are T-G, G-T, C-A, and A-C. A-A mismatching reduces amplification efficiency about 20-fold, while A-G, G-A, and C-C mismatches reduce efficiency about 100-fold compared with amplification from primers correctly matched to their template sequences.

Sequence complementarity is most important at the 3′ end of the primer as it is this end of the primer that is most important for specificity and extension. Primer sequence is crucial if there is no step following amplification and the genotyping relies on recognition of SNP alleles by the primers, as in allele-specific oligonucleotide (ASO) PCR, where primers are designed to amplify only if they match the alleles. Such genotyping schemes depend on exact complementarity between the primer's 3′ end nucleotide and its target nucleotide (**Figure 7.2**).

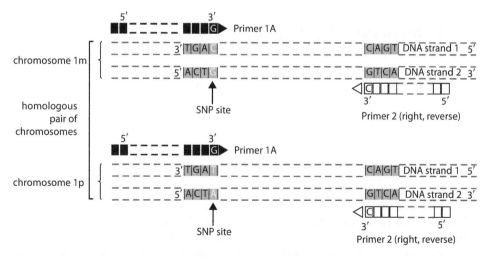

Figure 7.2 Primer locations for a genotyping scheme to determine a SNP by the presence or absence of amplification. The pair of forward primers has to be designed so that their 3′ ends coincide precisely with the SNP. In this example, the SNP is a C > T substitution: primer 1A is designed to anneal to the C allele on one of the chromosomes, and primer 1B anneals to the T allele on the other chromosome. The presence or absence of each allele can be inferred by amplification or its absence in separate reactions. Primer 2 is the same for each reaction.

Large gene families pose a problem for primer design

In practice, another issue that may be encountered is designing primers for genotyping a SNP in a gene that is a member of a large gene family. Heat shock protein genes are an example, with more than 20 members of the family, all of which have highly conserved (and therefore similar) sequences. Any primer pair designed to amplify a fragment in one gene in the family will inadvertently amplify similar fragments in other members of the family. This is the most common reason for deviations in Hardy–Weinberg equilibrium (HWE), which result from unintended amplifications of **paralog sequence variants** (**PSV**) or **pseudoSNPs**. An important implication of this situation is that such SNPs cannot be included in microarray-based genotyping platforms that are used routinely in genome-wide association studies (GWASs), and it reduces the coverage of certain areas in the genome.

Highly polymorphic genome regions are also difficult for primer design

With increasing knowledge of the human genome sequence, another problem in primer design has become clear. In highly polymorphic regions of the genome, such as the HLA region on chromosome 6p21.3, finding nonpolymorphic regions at which to place the primers may be very hard. If the primers contain polymorphic positions, amplification efficiency may be affected, depending on the proximity of the polymorphic nucleotide to the 3′ end of the primer. If the polymorphism is known, it may be possible to overcome the issue by designing **degenerate primers**. These are a mixture of oligonucleotides with different nucleotides in the positions that are polymorphic in both alleles of the SNP. While this seems to be a solution, it is not an ideal one because of complications if there is more than one polymorphism within the primer sequence or if the polymorphism has three or four alleles. In the latter case, a primer can be synthesized with the nucleoside analog inosine, which hybridizes to adenine, thymine, or cytosine, in order to prevent the polymorphism disrupting primer annealing.

Amplicon size can also be important

The size of the amplified fragment (the amplicon) is important in real-time PCR-based methods such as the TaqMan® allelic discrimination assay and high-resolution melting analysis. The amplicon size should be as short as possible (<150 base pairs [bp]; typically 75–100 bp) for successful genotyping. If the SNP is in a GC-rich area or within a repeat sequence, designing good primers will also be difficult, and their target sequence may have to be chosen outside of those difficult regions.

Oligonucleotide melting temperature is used in genotyping

Each oligonucleotide has a **melting temperature** (**Tm**) that is determined primarily by its nucleotide sequence. Nucleotides C and G contribute more than A and T to the melting temperature. Even a single nucleotide change may alter the melting temperature. The alteration in melting temperature due to a single nucleotide is more easily detectable in a shorter fragment. High-resolution melting-curve analysis is based on detecting the difference in melting temperature caused by a nucleotide substitution. For this method to be successful, the primers should be designed to amplify a very short fragment (<150 bp). Templates rich in GC nucleotides are more difficult to amplify due to stronger bonds between these nucleotides. Dealing with such templates requires special attention and additional chemical modifications.

Probes may also be necessary for genotyping

While some older genotyping methods rely solely on amplification, more modern methods require additional oligonucleotides (probes) to match the alleles of the polymorphism and its flanking regions (**Figure 7.3**). During DNA replication, nucleotides are added to the 3′ end of the newly forming strand due to the 5′ → 3′ polymerase activity of DNA polymerase. This is especially important to remember in the design of the real-time PCR-based TaqMan assay in order to select the strand to which the probe will anneal. The 5′ → 3′ exonuclease activity of *Taq* polymerase degrades the 5′ end of an annealed probe during primer extension toward that end of the annealed probe. It is therefore crucial that the label to help with detection of amplification is attached to the correct end (the 5′ end) of the probe.

Algorithms are used to help design effective primers and probes

The considerations to primer and probe design given here are not all-inclusive, and there are many different scenarios that impact on the success of primer and probe design. Much can be learned with experience, but ultimately it is **primer/probe design algorithms** and high-quality sequence information on regions of interest that contribute most to successful design. Additional considerations, especially when probes are involved in genotyping, should be the compatibility between probes, and between each probe and the primers. Primer design algorithms incorporate these considerations into the design process and assist with the design of the most suitable probe as well as primers. These algorithms have been converted to freely accessible online tools, which have become close associates of any molecular geneticist in their daily life. Several of these online tools are listed at the end of this chapter.

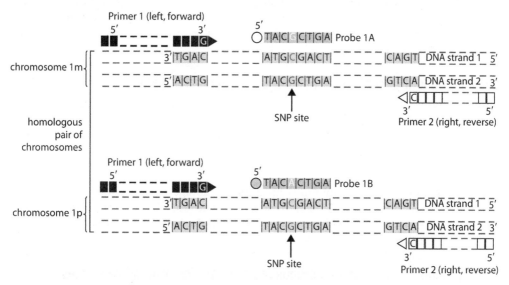

Figure 7.3 Probes used in genotyping. Probes are designed to have a perfect match at every nucleotide in the target sequence, with one probe for each allele of the SNP on different chromosomes. In this example, the SNP is a C > T substitution, and the two probes have G (probe 1A) or A (probe 1B) at the corresponding sites to anneal to the SNP alleles C and T, respectively. By attaching different fluorescent labels (stars) to the 5′ ends of these two probes, the presence or absence of each allele can be inferred by measuring the fluorescent emission. In this example, the sample has the CC genotype and only probe 1A will anneal to its target.

7.3 Choice of Genotyping Method

Rapid developments in molecular genetics have resulted in drastic changes in commonly used genotyping methods. Today, hardly any radioactivity is used for genotyping. The amount of DNA needed to get results has reduced from micrograms (required for Southern blotting-based assays) to picograms (required for real-time PCR-based assays), and now a total of only 200 nanograms is needed for genome-wide coverage using microarrays. Likewise, the time to results has reduced from days to minutes. With these new developments, many older methods have become almost obsolete (for example, Southern blotting-based genotyping is very rarely used), and highly sophisticated new methods have emerged.

As shown in **Table 7.1**, one of the considerations for the choice of genotyping method is the number of samples to be genotyped and the number of genotypes to be generated. At one end of the spectrum, small candidate gene studies involve a small number of samples to be genotyped for not many SNPs (low-throughput genotyping). The other extreme is a very large, multicenter GWAS that uses tens of thousands of samples to be genotyped for millions of SNPs (high-throughput genotyping). Low-throughput methods that are usually performed manually are not suitable for a GWAS, and using a high-throughput method that requires expensive instruments is equally unsuitable for a low-throughput study.

Ultimately, the choice of genotyping method is influenced by the size of the study, its purpose, and available resources. Cost, degree of throughput, resolution needed, and availability of specialty equipment will determine the choice of genotyping method. The cost of genotyping has come down considerably in recent years, the number of core laboratories in larger institutions has increased, and many companies now offer affordable genotyping, which allows it to be outsourced. An additional advantage of outsourcing genotyping to core laboratories or commercial companies is that the genotyping is performed by specially trained people using highly validated methods on sophisticated instruments, usually including robotics. In these purpose-built and well-resourced environments, turnover time is short and genotyping errors are minimized.

Low-throughput genotyping is still used in certain circumstances

Table 7.2 outlines some of the well-known low-throughput genotyping methods. It needs to be recognized that even GWAS results may need to be complemented by alternative genotyping methods, especially for parts of the genome where polymorphisms cannot be included in microarrays due to extreme polymorphisms, or where high-throughput methods cannot be used due to pseudoSNPs in paralog genes.

Low-throughput genotyping methods have the advantage of not requiring specialized instruments, but they have a high genotyping error rate, are labor intensive, and are risky in that amplified products need to be handled within the laboratory to obtain genotypes, which increases contamination risks.

High-throughput genotyping methods are required for large-scale studies

Low-throughput methods are not a viable option for large-scale genetic association studies. The most popular high-throughput genotyping methods are shown in **Table 7.3**. They are all closed-tube methods, meaning that the contamination risk is minimized because the tubes that contain PCR-amplified fragments are never opened within the laboratory. In fact, most of these methods use 96- or 384-well microwell plates that are sealed after the reaction setup and never opened.

Table 7.2 Low-throughput genotyping methods suitable for small-scale studies

Method	Features	Pros and cons
Restriction fragment length polymorphism (RFLP) typing	Relies on the availability of a restriction endonuclease enzyme that is able to cut a short sequence when one allele of the SNP is present, and not when the other allele is present	RFLP typing requires handling of PCR products and carries the potential risk of DNA contamination. It can only be used with extreme care, and for small-scale studies. Primer design is the least constrained (they can target any sequence outside the vicinity of the SNP)
Allele-specific oligonucleotide (ASO) PCR	Two separate reactions are set up with separate primers specifically designed to amplify the fragments carrying the alleles of a SNP. Amplification indicates the presence of the allele	ASO typing also carries the potential risk of DNA contamination. The need to set up two reactions per sample is a disadvantage. Primer design is constrained (their target sequence is the variant alleles and flanking regions). Most suitable for class I and II SNPs, and is not recommended for class IV SNPs. The detection of amplification on a gel is now largely replaced by detection using DNA dyes in a real-time PCR instrument
Sequence-specific oligonucleotide probe (SSOP) typing	Following PCR amplification, alleles are detected by two probes specific for each allele of the SNP. The binding of probes to the specific allele is detected by a biochemical reaction	Primer design is not constrained (they can target any sequence outside the vicinity of the SNP). The SSOP method has been largely replaced by the TaqMan allelic discrimination assay (in which detection occurs during the end of the PCR amplification without further handling)
Single-strand conformation polymorphism (SSCP) typing	Based on the different mobilities of DNA strands containing single nucleotide differences when run on a denaturing polyacrylamide gel. Used to be useful for screening a fragment for new polymorphisms	The amplified fragment containing the SNP should be less than 300 bp long and should not contain another SNP. SSCP typing has been largely replaced by melting-curve analysis, which is a closed-tube system that does not include opening of a tube that contains PCR products

The methods listed in **Table 7.3** require specialized instruments. They are either real-time PCR-based methods, which can be performed on relatively affordable equipment available in most laboratories, or require mass spectrometry, which is usually available only in core facilities. Another feature of these methods is that they need a very small amount of DNA for genotyping. Mass spectrometry, TaqMan, and melting-curve analyses can be achieved using picogram amounts of DNA per genotype. Microarray genotyping is capable up providing up to 5 million genotyping results by using around 250 ng of DNA (the smallest amount of DNA per genotype).

The TaqMan allelic discrimination assay, genotyping by melting-curve analysis, and the MassARRAY® iPLEX® assay are currently the most commonly used methods for analyzing up to 1000 SNPs in a candidate gene study. Of these, the TaqMan and melting-curve assays can be run in a modest laboratory which has a real-time PCR instrument, whereas iPLEX requires a specialized instrument.

GWASs use microarrays for genotyping

Microarrays are commercially developed products that incorporate the genotyping of up to five million SNPs. Microarrays are now the highest-throughput platform and provide the broadest coverage of human genome variation. Most core facilities have the necessary

Table 7.3 Selected modern genotyping methods suitable for larger studies and a large number of genotypings

Method	Features	Pros and cons
TaqMan allelic discrimination assay (5′ fluorogenic assay; hydrolysis assay probe)	Uses fluorescently labeled probes to detect the presence of each allele. Two primers to amplify a short fragment and two labeled probes are used	Can be run in any laboratory that has a real-time PCR instrument. Uses a minimal amount (picograms) of DNA. Can also be used to genotype copy number variations. Assays for most SNPs are commercially available
Melting-curve analysis by high-resolution melting (HRM) analysis	Relies on changes in the melting curve of short PCR amplification products caused by differences in their nucleotide composition. For most SNPs, only two primers are usually sufficient for genotyping; some SNPs may require a simple, unlabeled probe to increase sensitivity	Can be run in any laboratory that has a real-time PCR instrument or a dedicated melting-curve analysis instrument. Uses a minimal amount (picograms) of DNA. Can also be used to genotype copy number variations. Assays for most SNPs are commercially available
Single base extension (SBE)-based methods	Using a special chemistry and a primer that binds up to the nucleotide immediately before the SNP allele, the single nucleotide then added to the primer is detected. Detection may be based on fluorescence or mass spectrometry. Pyrosequencing, mass spectrometric genotyping (iPLEX®), and Illumina (Infinium®) assays use this approach	These methods require specialized instruments and are usually performed in core laboratories. iPLEX is able to genotype multiple SNPs (up to 40) in a single assay

equipment to run these assays, and when outsourcing is possible, microarray genotyping is both affordable and very reliable. For microarray genotyping, around 250 ng of high-quality genomic DNA is needed. Ideally, the DNA should be native DNA that has not been subject to **whole-genome amplification** (WGA), and all samples should be prepared by the same DNA extraction methods. Besides copy number variation typing, microarrays can also analyze cytogenetic changes. Given the high error rates associated with sequencing at low depths of coverage, microarrays containing validated SNPs provide extensive information on truly polymorphic variants. They are currently the highest-throughput platforms for genome-wide genotyping until next-generation sequencing becomes more affordable and accessible.

Next-generation sequencing is the future

Next-generation sequencing (NGS) is increasingly becoming the method of choice for truly genome-wide association studies, especially when rare variants are of interest, but only for researchers who have the resources to use it. NGS involves massively parallel DNA sequencing reactions (which do not use the traditional Sanger dideoxy chain-termination method) and casts the widest net to identify susceptibility variants in any disease. It is most useful in the identification of single mutations in Mendelian disorders. NGS is not a method that can be set up, run, and analyzed by a single individual in their own laboratory. A large team of experts, in particular bioinformaticians, and a large capital investment are required. Currently, NGS is only performed in well-resourced, specialized centers.

Next-generation (massively parallel) sequencing involves the parallel sequencing of up to millions of small fragments, and the sequence information from each is then put together using bioinformatics to generate the whole genome sequence. A major advantage of NGS is that it does not rely on prior sequence information, and it sequences any genome however different it may be from the reference sequence. Because the same fragments are also sequenced in parallel many times, it also provides more reliable sequences. Another advantage is that a single-cell genome can be sequenced, and this can be achieved without any pre-sequencing PCR step, thus avoiding any errors that may be introduced by PCR amplification. While it is technically possible to sequence the whole genome, it is more common that only exomes (which make up no more than 2% of the genome) are sequenced. In the context of genetic associations, exome studies are increasingly used to identify rare missense mutations causing diseases. Exome studies have already been used with success in both monogenic and complex diseases to discover rare mutations, and they will be used with increasing frequency until whole-genome studies become affordable and more practical. When this happens, truly genome-wide studies with the highest possible resolution will be achievable; at present, microarrays include only known SNPs and only those that can be genotyped in a microarray. With next-generation sequencing, there is no limit on the number and nature of rare and common variants that can be genotyped.

7.4 Sources of DNA for Genotyping

For successful genotyping, the nucleic acid template should be of the highest quality and purity. It is also important that the quality of samples from two comparison groups (usually cases and controls) should be comparable. In a genetic study of germ-line DNA sequence variants, DNA for genotyping may be taken from any cell with a nucleus (that is, nuclear DNA). The source of DNA could be peripheral blood or bone marrow white cells (red cells do not have nuclei), saliva, mouthwash, or buccal swab, or tissue samples (fresh, frozen, or embedded). Samples from fossils or museum material (for example, mummies) have been used for genetic studies, and a variety of samples (cells left on a toothbrush or a cigarette stub, hair from a brush, or blood stains on clothes, for example) have been genotyped for forensic studies. It is not practical to use such samples for large genetic epidemiologic studies. The ideal situation is where all participants in a study have provided biological specimens of the same kind and high-quality DNA has been extracted by the same method.

In countries where neonatal screening is routinely done, and the blood spots collected on Guthrie cards are stored, these blood samples from newborns provide another source of DNA for population studies. DNA extracted from these spots is generally of good quality, although storage conditions can lead to some degree of degradation and fragmentation of DNA. The implications of such DNA damage are that methods based on short PCR fragments, such as microarrays and TaqMan chemistry-based methods, are preferable for genotyping. Only a small quantity of DNA can be extracted from neonatal blood spots; thus, if a large number of genotypings are planned, it is necessary to use WGA.

WGA is useful when there is little DNA available for genotyping

Whole-genome amplification is a well-established method to increase the amount of DNA available for assay from as little as 10 ng of genomic DNA. Different companies have different methods to make copies of whole genomes without preferential amplification of certain regions. Some methods are suitable for generating long DNA fragments, and some are good for generating small fragments. It is therefore important to check the features of each method and to make the right choice of WGA method for the purpose. It is generally agreed that whole-genome amplified DNA is as good as native DNA for SNP genotyping by almost any method. However, some microarray methods require WGA as the first step of sample processing and so it may not be a good idea to use microarray genotyping if the samples have already been subject to WGA. Again, it is best to check the suitability of whole-genome amplified DNA for the particular microarray being used before proceeding. While it is safe to use whole-genome amplified samples for SNP typing, they should not be used for repeat polymorphism typing, since these repeats may have changed during the amplification process.

LCLs provide DNA that can be used for genotyping

Lymphoblastoid cell lines (LCLs) are B lymphocytes that are transformed by Epstein–Barr virus (EBV) to continue proliferating and provide an unlimited source of DNA. The **HapMap Project** samples, for example, are all EBV-transformed LCLs. LCLs have been widely assessed for the validity of genotyping results obtained from them, and it is generally accepted that LCLs provide SNP genotyping results equivalent to those obtained from native DNA samples. The same conclusion has also been reached for gene expression studies, but not for epigenetic studies. It is therefore acceptable to use LCLs for genotyping studies.

7.5 Genotyping Errors

Whatever genotyping method is used, and whether or not a human or software determined the genotype, genotyping is subject to error. Since errors cannot be avoided altogether, the aim is to minimize the error rate and reduce bias in final conclusions. It is unrealistic to expect a 0% error rate and keeping this rate below 1% is the aim. Various surveys have concluded that the error rates range between 0.5% and 30% in published studies. Even in high-throughput centers, an error rate of 0.5% is observed for duplicate genotyping of the same samples. It has been documented that even an error rate as low as 0.5 to 1.0% has the potential to obscure important findings. Currently, the smallest error rate is observed from using microarray chips in GWASs, but this is largely due to preselection of polymorphisms that are least likely to cause error. Genotyping errors can be due to features of the DNA sequence flanking the variant, a poor quality and quantity of DNA, the source of DNA, and the precision of the instrument used in genotyping. However, the most important source of genotyping error is human error in handling the experiments and data. Believing that humans are the greatest generators of error is a healthy approach to reduce genotyping error rates.

Genotyping errors have serious consequences

Genotyping errors are one of the reasons for a failure to replicate first observations and are often underestimated. The consequences of genotyping errors include finding spurious associations (false positives), which may never be replicated elsewhere, and missing existing associations (false negatives) due to errors equally distributed in comparison groups. Nondifferential (random) errors in comparison groups reduce the statistical power to detect an association: the sample size needs to be increased by up to 8% for each 1% rise in random genotyping error in order to maintain constant type I and type II statistical error rates. Differential (nonrandom) errors (differing error rates in comparison groups) have a more drastic impact on results than nondifferential errors. Given the expected odds ratio range of 1.1 to 1.5 in complex disorders, it is easy to imagine how errors can cause false-positive results. Common causes of differential errors are usually related to differences in the DNA source, collection and extraction methods, and storage and transfer to the genotyping center. Differences in DNA quality occur if cases are recruited in one center and controls in another, or already existing controls have been used. When such differences are present in samples for each comparison group (cases and controls, for example), even blinded genotyping may not remedy the problem of nonrandom genotyping errors, which may result in either spurious associations or false-negative conclusions. Errors in microsatellite typing may contribute to inflated false-positive rates among reported associations in family-based studies.

Genotyping errors can be minimized with careful planning and execution

There are several ways to minimize errors in genotyping at the study design phase and genotyping phase, and there are statistical approaches to examine potential errors at the analysis phase (**Figure 7.4**). The latter is most useful for errors that cannot be avoided easily. Efforts to avoid errors should be made at the initial phase of the study, before genotyping begins. Ideally, all genotyping must be performed by the same method, in the same laboratory, by the same personnel. All cases and controls should be recruited prospectively, the same type of biological specimens should be collected from all of them, and the same DNA extraction method and genotyping scheme should be used. Problems may arise if an already genotyped control group is being used and cases are genotyped by a different method. Likewise, if case and control groups differ in the DNA extraction methods or sources of DNA, this may lead to unexpected differences in genotyping results. If such differences in the source of DNA, extraction methods, or storage conditions cannot be avoided, preliminary studies should be conducted to show that any differences in samples (peripheral blood versus buccal swab, for example), DNA extraction methods (salting out versus commercially available kits), or the way DNA was suspended (just water or a buffer) or stored (frozen with no thawing, repeated freezing–thawing, or refrigerator only) do not cause genotyping errors.

Genotyping errors can occur for many other reasons including unknown alleles in the locus (such as tri-allelic SNPs), errors in assay design (for example, polymorphisms within primer-binding sites), errors in assay conditions (such as mistakes in thermal cycling

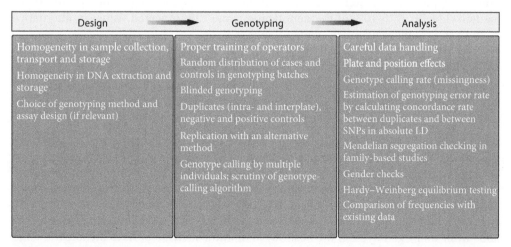

Figure 7.4 Measures to minimize and detect genotyping errors in different phases of a genetic association study.

parameters or PCR chemistry), errors in assignment of genotypes (genotype calling), errors in reporting of results, and errors in entering results into data sets. Of these, the most controllable are human errors, which include errors in the labeling of samples, leading to the mixing-up of samples; mishandling of biological specimens and DNA samples that lowers sample quality; contamination of samples, resulting in excess heterozygosity; errors in genotyping assays and genotype calling; and errors in data entry. Most of these can be minimized with proper training of operators and sound laboratory protocols with efficient supervision and automation of DNA extraction, liquid handling, genotype calling, and data transfer whenever possible.

Genotyping errors can be minimized at the experimental phase

Blind genotyping is one approach that can be adopted to reduce genotyping errors. In this approach, the operator doing the genotyping is unaware of the samples' identity in terms of which comparison group they belong to. It is also important that each batch for genotyping should include samples from all comparison groups. In other words, it is not a good idea to genotype cases first followed by controls. In each batch, there should be duplicates unknown to the operator and they should yield the same results. These controls are called intraplate duplicates when a microwell plate is used for genotyping. Ideally, the blind duplicate concordance rate should be 100% or close to it. When more than one plate is used for genotyping, each plate should have a blind replicate (interplate control) that is expected to yield an identical result. The total duplication rate should be 5 to 10% of the total number of samples. This practice allows an accurate estimation of genotyping error rate and the necessary correction of statistical results.

As in any scientific experiment, genotyping experiments should also have negative controls (for example, controls that have no template in TaqMan assays) and positive controls (inclusion of samples with known genotypes or the same samples in each batch). Interpretation of the results from these controls is straightforward and helps assess the reliability of the genotyping, including the human operator's reliability. Samples with known genotypes such as HapMap Project samples can be obtained commercially (for example, from the nonprofit Coriell Institute for Medical Research).

Another approach is to replicate a proportion of the genotyping using an alternative method. Ideally, 5 to 10% of the genotyping should be repeated using an alternative method and should yield a 100% concordance rate.

Genotyping errors can be assessed at the analysis stage

While the measures described above are used at the planning and experimental phases of a study, once the study results are obtained there are still a few more checks required to assess genotyping errors, even if all prior checks have been satisfactory.

There may be plate-to-plate variations in sample quality, interrelatedness among samples, and experimental conditions that may influence the genotype-calling algorithms of different vendors. In TaqMan genotyping, some real-time PCR instruments may generate variable results depending on the position of the samples in the plates, especially for those on the edges of plates. This edge effect may also be due to evaporation from wells on the edge resulting from not sealing the plate well. Researchers should be aware of such situations while doing genotyping. The best way to monitor such factors is to include the same internal control samples in each run, and to check whether they produce the same genotype every time.

Ideally, all samples that have been included in the genotyping effort should yield a geno-type. The proportion of samples that do not produce a result makes up the missing-call rate. The aim is to obtain results from all samples (100% genotyping rate). Lower rates result from low sample quality or choosing an unsuitable method of genotyping. When genotype call rates are lower than a predetermined threshold (lower than 95% is unacceptable) it becomes important to assess whether the samples with missing genotypes are randomly distributed across comparison groups. Differential rates of missing calls (or "missingness") between cases and controls, despite random distribution in genotyping batches (like plates), may be a problem that generates a bias that distorts the results. Another problem with genotype calling is the variation between observers. Observer variability should be minimized by double-blind scoring and data entry, which should be followed by electronic comparison of the blind entries for concordance.

Examination of **Mendelian segregation** is the most reliable mechanism to check for genotyping error but can only be used in family-based studies. When genotype data are available from parents and their offspring, it is easy to examine the fit of the data to Mendelian expectations. Any unexpected result that does not conform to Mendelian segregation patterns (such as finding heterozygote offspring of parents who are both homozygote for the same allele) points toward genotyping errors. Having this internal genotyping quality control mechanism is an advantage of family-based study designs over case-control studies. However, when genetic markers are exclusively SNPs, only around 25% of random genotyping errors will be detected via this method in a parent–child triad design. The probability increases when the allele frequencies of a SNP are near equal, and decreases when the variant allele is rare. Another threat to the successful use of Mendelian error checking is the incorrect assumption of paternity. The rates of nonpaternity are higher than expected, especially in Westernized societies.

If data from X- and Y-chromosome SNPs are available, results from these SNPs should be compatible with the gender of the samples. Female samples should have no genotype for Y-chromosome SNPs and male samples should not be heterozygote for X-chromosome SNPs. The data should also be checked for consistency in allele frequencies across plates. The frequencies in controls should also be similar to the frequencies already known for the population examined. If there are pairs of SNPs in absolute correlation (or perfect linkage disequilibrium), they should yield identical genotypes. SNPs with pairwise $r^2 = 1$ in all four HapMap phase I panels are expected to be in perfect LD in any population, and these can be used to estimate genotyping error rate. Allele shifting is a typical error that occurs in microsatellite typing when typing is done in batches, but this problem does not apply to SNP typing.

There is also specialized software that can check for, and quantify, the magnitude of genotyping errors. Most software packages, however, work in family-based studies and rely on Mendelian segregation. Others check HWE or replicates, or run simulation and statistical power calculations to estimate the impact of errors on association results. Some algorithms run association statistics allowing an underlying error model.

HWE testing is frequently misused

HWE testing is another check routinely performed at the analysis stage of an association study. HWE violation may be due to genotype errors, but there are other reasons too. As discussed in earlier chapters, HWE assumes random mating, lack of selection according to genotype, and absence of mutation or gene flow. It is unclear whether these assumptions are always met in human studies and to what degree they may be responsible for violation of HWE. It is generally assumed that an outbred human population can be thought free of those factors that violate HWE. Therefore, HWE testing is commonly used to rule out genotyping errors, but this method may not be as reliable as is thought. If HWE testing shows violation in controls but not in cases, it is more likely that genotyping errors have occurred. It has to be remembered, though, that HWE testing can only detect gross genotyping errors because the statistical test for HWE has a very low statistical power. Unless the sample size is very large, no mild or moderate rate of errors will result in statistically detectable violation of HWE. Thus, while a violated HWE indicates a significant error rate, no violation of HWE is no guarantee for a lack of errors.

What if HWE is really violated?

If violation of HWE is not due to genotyping errors—as shown by the same results being obtained by use of an alternative method—then, in the absence of obvious biological reasons, several possibilities should be considered. One potential reason is an unknown allele in the locus being examined, which will be missed by the genotyping assay. Samples heterozygous for the unknown allele will appear homozygous for one of the known alleles and, if present, homozygotes for the unknown allele will not reveal any genotype. Unknown polymorphisms within primer annealing sites may cause nonamplification of the allele on that chromosome. This situation is similar to that of an unknown allele, but is actually missing a known allele due to primer malfunction. In the cases of unknown alleles and polymorphisms within primer sequences, the result is a deficit in heterozygosity due to allelic dropout, which is the most prevalent form of HWE violation. The inbreeding coefficient (FIS) is a measure of the degree of deficit in heterozygosity and deviation from HWE.

If the variant of interest is in a structural polymorphism region, there will be no result if the region is deleted. However, if the polymorphism is a CNV, and if the polymorphism is variable within each unit of the CNV, the results will violate HWE whatever genotyping method is used. Likewise, if the polymorphism is in a gene that has paralogs within the genome, a similar situation arises. Depending on whether the site is polymorphic in paralogs (pseudoSNPs) and which allele is present at the nonpolymorphic sites in paralogs, HWE may be violated despite the genotyping itself being accurate. In the case of structural polymorphisms and the presence of paralogs and pseudoSNPs, results are variable, but usually excess heterozygosity is observed. Violation of HWE due to pseudoSNPs is the easiest to detect by statistical tests, even in small samples. Another unrelated reason for excess heterozygosity is contamination of samples.

In the case of population substructure, even when genotyping is accurate, results will show violation of HWE. The deviation is usually due to increased homozygosity in the whole population sample that has cryptic substructure. While violation may be due to these reasons, the most common cause remains genotyping errors.

When results are found to violate HWE, further explorations are needed before discarding the results. Assuming no biological reasons can account for the violation, genotyping should be repeated using an alternative method. If results are still the same, they can only be analyzed using additive model analysis. If statistical assessment yields no indication of HWE violation, the power of the statistical test should be considered. The results from very large studies are valid only if they show no HWE violation. A review of the literature has revealed that SNPs with low minor allele frequencies are most likely to violate HWE due to genotyping errors. The interpretation of results from rare SNPs, therefore, requires attention.

7.6 Imputing Genotypes

Genotyping itself is no longer the only way to generate genotype results. Thanks to the known LD patterns within a population, results for any SNPs that have not been genotyped can be derived by computer algorithms as long as there is reference population data for all SNPs in the genome. This process is called **imputation**. The availability of whole-genome DNA sequence data from more than 1000 individuals in the 1000 Genomes (1KG) Project enables researchers to impute SNP data without genotyping SNPs. To be able to impute genotypes, a study should yield high-resolution genotyping results in the region of interest, and the rest of the SNPs (usually rare variants) identified in the 1KG project can then be imputed or inferred by recently developed statistical approaches that use the 1KG results as reference panels. This approach has been documented to yield reliable results, and more and more studies are adopting this strategy rather than actually genotyping millions of additional rare variants. Obviously, the reliability of the imputed genotypes correlates with the actual original genotyping in the region of interest. The imputation approach is also being used in meta-analysis of previous GWAS results by imputing missing genotypes in all GWAS sets before combining them for the final analysis. MACH (Markov Chain-based Haplotyper) software is one of the tools used for genotype imputation. Imputation methods have also been used for imputing HLA alleles from SNP data. HLA*IMP is one such algorithm. All imputation efforts exploit the correlation structure (LD) existing throughout the human genome that results in the ability to infer the genotype at a second locus once information at one locus is already available.

Key Points

- A good understanding of molecular biological principles, from nucleotide chemistries to genomic structural variations, is necessary for successful genotyping assay design and also to help with minimizing genotyping error rates.

- There are still many genetic association studies performed in smaller laboratories using low-throughput genotyping methods.

- Low-throughput genotyping methods are more likely to have a high genotyping error rate than commercially developed high-throughput methods.

- The genotyping error rate is usually underestimated and has the potential to distort results.

- The most common source of genotyping error is human error.

- HWE testing is not a reliable approach to detect genotyping errors.

- If HWE is violated with no biological reason such as selection or inbreeding, there may be genomic features that interfere with the genotyping process.

- Commercially designed and quality controlled microarrays are the most reliable genotyping method available, but they leave out SNPs from certain regions of the genome due to technical difficulties. Coverage of genome-wide SNPs should not be assumed to be 100%.

- The ultimate genotyping method is next-generation sequencing, but the cost and additional expertise needed to do it currently restrict its use to major centers.
- Imputation is a well-established method to generate more genotype data from existing data using specialized software.

URL List

Commonly used primer design algorithms

Primer3. Whitehead Institute for Biomedical Research. http://primer3.wi.mit.edu

PrimerQuest. Integrated DNA Technologies (IDT). http://www.idtdna.com/Primerquest

Some examples of companies offering custom genotyping solutions

Affymetrix. http://www.affymetrix.com

Coriell Institute for Medical Research. https://www.coriell.org/research-services/genotyping-microarray/overview

Illumina. http://www.illumina.com/techniques/popular-applications/genotyping/custom-genotyping.html

LGC Genomics. http://www.lgcgroup.com

Software for genetic error detection

Error Checking Programs. Section on Statistical Genetics (SSG). University of Alabama. http://www.soph.uab.edu/ssg/linkage/errorchecking

Further Reading

Genetic variant typing

Dorak MT (2007) Genotyping with PCR. How to choose the right approach. The Scientist 21, 70–72 (http://www.the-scientist.com/?articles.view/articleNo/25120/title/Genotyping-with-PCR).

Edenberg HJ & Liu Y (2009) Laboratory methods for high-throughput genotyping. *Cold Spring Harb Protoc* 2009(11), pdb.top62 (doi: 10.1101/pdb.top62).

Landegren U, Nilsson M & Kwok PY (1998) Reading bits of genetic information: methods for single-nucleotide polymorphism analysis. *Genome Res* 8, 769–776 (doi: 10.1101/gr.8.8.769).

Liu L, Li Y, Li S et al. (2012) Comparison of next-generation sequencing systems. *J Biomed Biotechnol* 2012, 251364 (doi: 10.1155/2012/251364).

Little J, Higgins JP, Ioannidis JP et al. (2009) STrengthening the REporting of Genetic Association Studies (STREGA) – an extension of the STROBE statement. *Genet Epidemiol* 33, 581–598 (doi: 10.1002/gepi.20410).

Ranade K, Chang MS, Ting CT et al. (2001) High-throughput genotyping with single nucleotide polymorphisms. *Genome Res* 11, 1262–1268 (doi: 10.1101/gr.157801).

Sobrino B, Brión M & Carracedo A (2005) SNPs in forensic genetics: a review on SNP typing methodologies. *Forensic Sci Int* 154, 181–194 (doi: 10.1016/j.forsciint.2004.10.020).

Taberlet P, Griffin S, Goossens B et al. (1996) Reliable genotyping of samples with very low DNA quantities using PCR. *Nucleic Acids Res* 24, 3189–3194 (doi: 10.1093/nar/24.16.3189). (*This study shows the difficulty of producing reliable genotype data with the occurrence of false alleles and false homozygotes [that is, allelic dropout].*)

Genotyping error

Akey JM, Zhang K, Xiong MM et al. (2001) The effect that genotyping errors have on the robustness of common linkage-disequilibrium measures. *Am J Hum Genet* 68, 1447–1456 (doi: 10.1086/320607). (*A study that investigates the effects of genotyping error on estimates of linkage disequilibrium; it shows that the robustness of the estimates depends on allelic frequencies and assumed error models.*)

Clayton DG, Walker NM, Smyth DJ et al. (2005) Population structure, differential bias and genomic control in a large-scale, case-control association study. *Nat Genet* 37, 1243–1246 (doi: 10.1038/ng1653).

Dequeker E, Ramsden S, Grody WW et al. (2001) Quality control in molecular genetic testing. *Nat Rev Genet* 2, 717–723 (doi: 10.1038/35088588).

Miller MB, Schwander K & Rao DC (2008) Genotyping errors and their impact on genetic analysis. *Adv Genet* 60, 141–152 (doi: 10.1016/S0065-2660(07)00406-3).

Pompanon F, Bonin A, Bellemain E & Taberlet P (2005) Genotyping errors: causes, consequences and solutions. *Nat Rev Genet* 6, 847–859 (doi: 10.1038/nrg1707).

Imputation

Ioannidis JP, Thomas G & Daly MJ (2009) Validating, augmenting and refining genome-wide association signals. *Nat Rev Genet* 10, 318–329 (doi: 10.1038/nrg2544). (*This review covers important topics pertinent to genetic association studies, including a not-too-technical explanation of imputation.*)

Candidate Gene Studies and Genome-Wide Association Studies

8

Genetic association studies use genetic variants. Given that there are millions of genetic variants in the genome, it is important to select carefully the variants to be used in association studies in order to obtain the correct answer to the research question and not to waste resources. Initial genetic association studies (candidate gene studies) only used one or a few variants from a candidate gene. Many contemporary genetic association studies are genome-wide association studies (GWASs), which use up to five million variants, but this is still not all the variants in the human genome. The ultimate aim is to encompass all variants, including previously unknown ones, which is only possible with whole-genome sequencing. In this sense, whole-genome sequencing can be viewed as the largest possible GWAS. It might be thought that candidate gene studies are now obsolete, but this is not the case and the place of candidate gene studies in the post-GWAS age will be discussed in this chapter. The advantages and disadvantages of GWASs will also be discussed. The chapter starts by describing the selection of candidate genes and variants, contrasting the requirements of candidate studies and GWASs.

8.1 Candidate Genes

While technological advances have enabled us to screen many genome variations without any previous knowledge, candidate gene studies need a starting point.

Candidate gene studies rely on linkage results or biological knowledge of the disease

If a genomic region is implicated in a linkage study, usually for a Mendelian disorder, it is identified as a candidate region for screening. The ΔF508 mutation in the *CFTR* gene, found to cause cystic fibrosis, was identified in studies following up from linkage studies. Another scenario is that a genomic region is implicated in animal studies and the corresponding region is then screened in humans. A good example of using previous animal studies is determining the involvement of the major histocompatibility complex (MHC) in leukemia development, which was first found in mice and then investigated by extensive studies in human MHC genes for associations with leukemia risk. The MHC is also an obvious candidate for autoimmune disease susceptibility loci and has been examined as a candidate region in autoimmune diseases. While these studies are all called candidate gene studies by virtue of not being GWASs, the candidate may actually be a region, a physiological pathway, or even a single polymorphism.

The most valid and unquestionable indication to do a candidate gene study is the statistically significant results of a linkage study that identifies a genomic region, typically 10–20 Mb in size, for the presence of a disease gene. The region identified may be large enough to harbor more than a hundred genes, depending on the gene density. When the number of genes in a linkage region is large, there may be multiple biologically plausible candidate genes that need further examination. Trying to narrow down the genomic region by increasing the density of markers in a linkage study may be successful, especially in Mendelian disorders. In cancer studies, examination of loss of heterozygosity in tumor cells from heterozygous individuals may nominate candidate regions that may harbor tumor suppressor genes.

Most of the past success in identification of the disease genes in Mendelian disorders, such as *BRCA1* in breast cancer and the genes involved in cystic fibrosis and chronic granulomatous disease, came from linkage studies. Nowadays, linkage studies are rarely performed. The reason for this is the affordability of GWASs, which allows skipping of the linkage study step in the search for a genetic basis of a disease. Even in the GWAS era, however, linkage studies are not yet dead. Linkage studies provide greater statistical power than association studies to detect genomic regions harboring **rare variants** that modify disease development risk with high penetrance. This is particularly true when multiple rare mutations in the same gene are involved in disease etiology. In cystic fibrosis, for example, there are more than 600 very rare mutations in the *CFTR* gene that can cause the disease and any one of them may contribute to the linkage results. Linkage studies do not require any prior knowledge of the gene function or possible location and are not susceptible to the effects of population structure, but they do require large multigenerational families with biological samples from family members.

Candidate gene studies may be exploratory

A gene can be hypothesized to be involved in the development of a disease simply because of its function—say, any apoptosis-related gene in cancer—but without any specific prior data. The candidate gene study based on this hypothesis is an exploratory study. Such studies are perhaps the most common variety of candidate gene studies, but rather than providing an answer to the research question, they often generate more questions to pursue in follow-up studies. There is nothing wrong with this, but care must be taken to avoid what are essentially "fishing expeditions" that chase statistical significance for publication purposes without much interest in disease biology. For example, if no overall association is found, but an association is indicated only in a subset of the people with the disease, further subgroup analyses are carried out until a statistically significant association is found that is judged sufficient to generate a new hypothesis to be explored by future studies. These associations are unlikely to be replicated in future studies, and subgroup analysis without a prior hypothesis is rarely acceptable.

Experimental evidence is needed to establish candidate gene status

For a gene to be declared a biological candidate, its function should be well known, that function should be related to disease development, and there should be direct evidence for this connection. With increasing knowledge of the multiple functions of each gene, as well as the multiple pathways involved in disease biology, it is not difficult to make connections between diseases and genes, but to establish a true candidate gene there should be experimental evidence that the gene is involved in disease etiology. For example, any apoptosis-related gene may be considered as a candidate for any cancer, but if the gene in question is not expressed in the tissue from which the cancer derives, it cannot be a candidate gene. Studies that retrospectively try to fit the results of an exploratory study with a role for that gene in disease pathogenesis can be recognized, and the **biological plausibility** of their results often appears questionable.

8.2 Candidate Variants

After selection of a gene, candidate gene studies then need to select a small number of variants to be genotyped. GWASs also select variants to genotype but they use a vast number of variants and so the selection criteria are different.

How to select the most informative variants for a candidate study

The earliest association studies used whatever few variants were known: blood groups (A, B, AB. or O) or genetically determined variants in enzyme activity (slow or fast acetylators or responders; poor metabolizers). Today, however, there are millions of known genetic variants and an informative subset has to be selected for a study. Even using a small gene that has a limited number of variants may waste resources (time, biologic samples, reagents, and money) if variants that do not provide additional information are genotyped.

Genetic associations result from alterations of gene function by variants. The first aim in selection is to include functional variants, although a nonfunctional variant may still be a proxy for an unknown rare functional one. **Figure 8.1** shows the three established approaches for the most effective variant selection: empirical, bioinformatic, and mathematical. Variants may already be known to be functional because of experimental evidence or may be predicted to be functional using bioinformatic approaches. The mathematical approach examines correlations among variants and those showing strong correlations with already selected functional variants are eliminated. If no

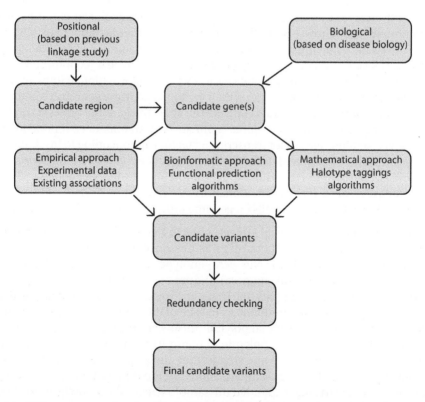

Figure 8.1. Different approaches in effective variant selection for candidate gene studies.

functional information is available, correlations among variants are still scrutinized so as not to leave any correlated pair in the final set of candidate variants.

The first step in candidate variant selection is assessment of the polymorphic content

In practice, variant selection begins with finding out how many variants are present in the gene(s) of interest. This information is more important for candidate gene studies in order to select the correct variants of the correct gene to include in the study. The polymorphic content can be assessed in a number of ways. NCBI ENTREZ SNP, commonly known as dbSNP, lists all known variants in any gene and also anywhere in between genes. Not all variants listed in this database have been confirmed as variations in a population study. Some variants listed have been reported to this database as a result of sequencing studies, and a fraction of those are sequencing artifacts. Such entries are easy to recognize by checking the validation information; if there is no confirmation shown by a population study, and they are found to be an artifact, they may be deleted. Some SNPs with no population data may be found to be highly deleterious to protein structure or function by bioinformatics analysis, and may be included in a study. Such SNPs may turn out to be nonpolymorphic. It is, therefore, a good idea to test the polymorphic nature of SNPs with no previous population data in a small pilot study. dbSNP provides information on genotype frequencies for SNPs if they have been examined in populations. This database also contains data regarding whether the SNP has shown any disease association, together with a link to the PubMed entry of the publication.

Genome browsers provide extensive information on each gene and its polymorphisms, together with many other features of the DNA sequence along each chromosome. The most commonly used browsers are the University of California Santa Cruz (UCSC) and Ensembl browsers. Using these browsers, a researcher can zoom in to or scroll over chromosomes to assess the location of the gene of interest and to see its genomic context, size, structure, polymorphic content, and splice variants. This information helps with selection of the most relevant variants by checking their location inside the gene and whether it varies in different isoforms. If a variant is outside the gene, the genome browser can be used to check whether it alters any functional motif such as an enhancer, a noncoding RNA gene sequence, or binding sites for transcription factors or noncoding RNAs. Increasingly, genome browsers incorporate data from other projects on genome bioinformatics. The UCSC browser, for example, is now linked to the Encyclopedia of DNA Elements (ENCODE) project database and provides information on the regulatory functions of SNPs as assessed by the ENCODE project. More details on genome browsers are provided in Chapter 10.

Phase I of the International HapMap Project aimed to generate a genome-wide database of SNPs to aid with the design and conduct of genetic association studies of common diseases. By the end of phase II, four million SNPs were genotyped in four major population samples (European, African, Chinese, and Japanese). The project was then expanded by genotyping seven additional population samples. The HapMap Project data provide insight on the polymorphisms in a gene and their frequencies in each of the major sample populations, and, most importantly, confirm the actual polymorphisms among a large proportion of those listed in dbSNP. Following the completion of phase III of the HapMap Project, the 1000 Genomes Project (1KG) has gone one step further and sequenced the whole genome of 1000 samples, including HapMap samples. This effort identified a lot of new, rare variants (frequency less than 0.01) and expanded the sample size for the analysis of existing polymorphisms. The 1KG data are incorporated in the Ensembl genome browser.

The empirical approach to selection of candidate variants is based on existing experimental data

While descriptive information on variants may be used to prioritize them for inclusion in a candidate gene study, any existing experimental data must be considered in candidate variant selection. Any variant that has shown an association with a phenotype or has been found to be functional experimentally has some value for a genetic association study. Previous associations, if validated by independent replication, may be with any phenotype and not necessarily with the same phenotype to be examined for genetic associations. The NIH Genetic Associations Database (GAD), NHGRI GWAS catalog, NHLBI Genome-wide Repository of Associations between SNPs and Phenotypes (GRASP), and NCBI database of Genotypes and Phenotypes (dbGaP) Association Results Browser list known genetic associations (see the URL list at the end of the chapter). These databases should be examined for known variant associations with traits as such associations contribute to the candidacy status of a variant.

Since variants usually modify disease susceptibility via their effects on gene expression, their expression quantitative trait locus (eQTL) status is an important factor in variant selection. Currently, the most promising resources for existing experimental information on functional SNPs are RegulomeDB and the GTEx (Genotype-Tissue Expression) project. RegulomeDB uses data from published reports, the ENCODE project, and the NCBI Gene Expression Omnibus (GEO) to annotate SNPs with known and predicted regulatory elements in the genome. The known and predicted regulatory DNA elements included are regions of DNase hypersensitivity, transcription factor binding sites, and promoter regions, especially in intergenic regions. Given that the amount of biologically active (transcribed) DNA is at least 15 times greater than the amount of DNA involved in making proteins, it makes sense to extend the search for functional variants beyond protein-coding genes. RegulomeDB is searchable for SNPs, genes, or genomic regions and ranks variants for their functionality based on empirical data.

The GTEx project database is a resource for the scientific community to study the relationship between genetic variation and gene expression in multiple human tissues. The samples used are freshly obtained autopsy samples, and genotyping and gene expression studies have been performed at high resolution using high-throughput methods. This project provides resources to correlate genetic variation with gene expression not in a single cell but in multiple cell or tissue types. No other project has analyzed genetic variation and expression in as many tissues in such a large population. The GTEx online tool is searchable by SNP ID, gene ID, or chromosome region. It also incorporates GWAS results for each phenotype examined. More information on bioinformatics tools is provided in Chapter 10.

The bioinformatics approach relies on computational prediction

Bioinformatics has developed as a major field in biomedical sciences in recent years, mainly for efficient data-mining purposes. Almost every issue of major bioinformatics journals includes reports on new algorithms for different aspects of genomic analysis. One of the goals of the bioinformatics field is the computational prediction of functional changes caused by genetic variants. This approach can be used either for selection of putative functional variants or to assess the functionality of variants found to be associated with a disease. There are many algorithms for the assessment of functionality and their details are beyond the scope of this book. However, searchable, user-friendly, online tools can be used for this purpose without much training in bioinformatics. There are tools for assessing the different ways a variant may alter gene function, such as considering the effects of amino acid substitutions on protein function or of alterations in transcription factor binding sites or splice sites. The most commonly used online tools are, however, bioinformatics suites that generate or compile all of this information in a single search. The most common of these include NCBI PheGenI, SNPnexus, and GWASdb. HuGE Navigator is another comprehensive suite that brings together in one place a number of tools useful for genetic association study design. The GenEpi Toolbox Website provides a good background to the scoring algorithms and various tools used to assess SNPs for different functionalities (see URL list).

Variant prioritization has changed since the advent of the GWAS. The emphasis used to be on **coding-region variants**, probably because most known disease-causing mutations were either nonsense or missense variants. Of all variant types, nonsense variants indeed have the strongest impact on gene function, followed by radical (as opposed to conservative) amino acid changes caused by missense variants. These types of variants are, however, very rare and their associations are very difficult to detect in candidate gene studies using small sample sizes. GWAS results have made it clear that variants with an effect on gene expression regulation, rather than those altering protein structures, are more relevant in common disease risk. **Figure 8.2** shows the different ways that variants can be involved in genome function depending on their location. Although the effect on disease risk of variants that affect gene expression is small, these variants have higher frequency in the genome and, as a result, most GWAS associations are from **noncoding regions**, including intergenic regions. Thus, the common variants associated with common disease risk are more likely from the noncoding regions and the rare variants are from coding regions. Since the effect size and frequency are both determinants of statistical power, and they are inversely correlated, variant selection is a matter of a trade-off. To be least affected by this trade-off and to eliminate chance effects, it is ideal to aim for a very large sample size in any genetic association study.

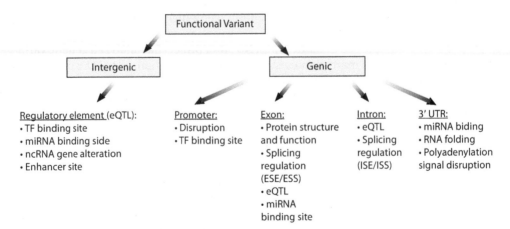

Figure 8.2. A functional variant influences gene activity via different mechanisms depending on its location. eQTL, expression quantitative trait locus; TF, transcription factor; miRNA, microRNA; ncRNA, noncoding RNA; ESE, exonic splicing enhancer; ESS, exonic splicing silencer; ISE, intronic splicing enhancer; ISS, intronic splicing silencer; UTR, untranslated region.

The mathematical approach does not take into account functions of variants

The third approach to variant selection relies on the fact that neighboring variants are often highly correlated or in high linkage disequilibrium (LD). This correlation can be exploited to select a few SNPs that also represent many others due to correlation. The subsets of SNPs selected using mathematical algorithms are called **tagSNPs**, which serve as proxies for a number of nearby SNPs in high LD. This approach does not use any information on functionality and SNPs are selected purely due to their ability to represent other SNPs in the region. The tagSNP approach aims to cover the study of a given genomic region by using the minimal number of SNPs to examine that region for association signals. If in the subsequent association study any signals are found, they are more likely to be proxies than to be the causal SNP. Further examination of the selected subregions for functional SNPs and their associations would then be the next step to identify causal variants.

The tagSNP selection can be carried out using various software packages. The freeware Haploview is the most popular package and was designed for the specific purpose of tagSNP selection using HapMap Project data. It is very user friendly and easy to learn using comprehensive user guides. Haploview is also capable of doing simple association analyses using case-control or family-based genotype data. The HapMap genome browser can also be used to screen genomic regions for selection of tagSNPs. In almost any genomic region, there will be some SNPs that do not show correlations with any tagSNP, and these SNPs should be included in a study alongside the tagSNPs. Using exclusively tagSNPs for the whole gene does not necessarily mean representation of all polymorphisms within the gene, and some other SNPs that cannot be tagged may have to be included for that purpose. The tagSNP approach is preferable over other approaches in a candidate study if a negative result will be taken to mean that common variants in a gene do not contribute to susceptibility.

Variant selection for a GWAS uses different approaches and has a different purpose

Although a GWAS covers as much of the whole genome as possible, the number of SNPs to be genotyped by a microarray chip should still be reduced to several million from the total number of validated SNPs known, which is currently in excess of 80 million. Custom chips may be designed to contain only nonsynonymous coding-region (missense) SNPs or only promoter-region SNPs, but default chips are not gene-centric and should cover both genic (coding) and nongenic (noncoding) regions of the genome. The two main approaches for SNP selection for all microarray chips are selecting SNPs spaced equally along each chromosome (random selection) and including tagSNPs for the whole genome. A combination of these two approaches, sometimes with additional SNPs selected for functionality from previous GWAS results, is also possible.

Given the ever-reducing cost of microarray production and genotyping, even candidate gene studies are now using custom microarrays more and more often. Microarray chips may contain from several hundred (candidate gene studies) to thousands (GWASs) of SNPs specifically selected for the purpose of the study. The most commonly used custom genotyping chip is the Immunochip, which consists of the SNPs in the HLA region and all other immunoregulatory genes, as well as SNPs anywhere in the genome that have shown associations with immune-mediated diseases. Other custom chips may cover SNPs in genes involved in certain pathways, like the nuclear factor κB (NF-κB) signaling pathway, or in a collection of genes that are known to interact in physiologic phenomena, such as iron regulatory genes. These custom chips usually require a minimum number for production and this should be taken into account in the design and costing of a study. Certain areas of the genome are also difficult to genotype on microarray chips and therefore cannot be included in a GWAS. These will be discussed in the next section.

8.3 Candidate Gene Studies in the GWAS age

In the past, every linkage study would have to be followed up by a candidate gene study. With the GWAS now being the main study design for genetic association studies, one might think that candidate gene studies are obsolete, but the GWAS has actually replaced linkage studies more than candidate gene (association) studies. This makes sense, as both GWASs and whole-genome linkage studies are hypothesis-free designs. While the most common follow-up study after a GWAS targets the rare variants in a specific gene or region, a follow-up study may focus on a gene or polymorphisms that could not be included in a GWAS. There are certain genes that require special attention in order to be genotyped and, for technical reasons, cannot be included in GWAS chips.

Candidate gene studies can be used after a GWAS to identify causal variants

Since early GWAS chips could not include more than 500,000 SNPs, a compromise was made to include just the relatively common variants; therefore, most historical GWAS results are unlikely to have identified causal variants (which are expected to be rarer) but rather established the proxies for causal variants. To identify the causal variants, candidate studies in targeted genes or regions are still used. The identification of rare causal variants can be achieved even after the study is completed by imputing rare variants from GWAS data, provided that the data contain sufficient SNP results to begin with, and that there are sufficient reference data for the population in question. With the exception of small isolated populations, reference data have been provided by the 1000 Genomes Project and imputation is possible for most studies.

By definition, rare variants have a very low frequency (<0.01) and their contribution to the overall inherited susceptibility to disease is very small, but individual odds ratios are very high. Reference data sets may not be large enough to contain all possible rare variants for imputation purposes, and sequencing may be the only option for selected regions with an association signal. The discovery of rare variants can be achieved by whole-genome sequencing in a very large case-control study (which is unrealistic due to the resources it requires) or by sequencing carefully selected candidate genes. Rare variants will usually act as a group to disrupt the function of a gene or a set of genes involved in the same pathway. For a candidate gene study designed to detect new variants, cases should be chosen carefully to be enriched for the presence of rare variants; that is, by including cases with one or more close relatives affected, and with an early age of onset for most disorders and especially for cancer.

Candidate studies are needed for large families of genes with high sequence similarity

Genes that belong to large families of closely related paralogous genes or those with a pseudogene that is the result of a duplication event are also particularly difficult to genotype using microarrays (**Figure 8.3**). A gene with a close relative showing a high degree of sequence similarity needs to be selectively amplified by taking advantage of the genomic sequences not shared by their close relatives. Only after this selective amplification by

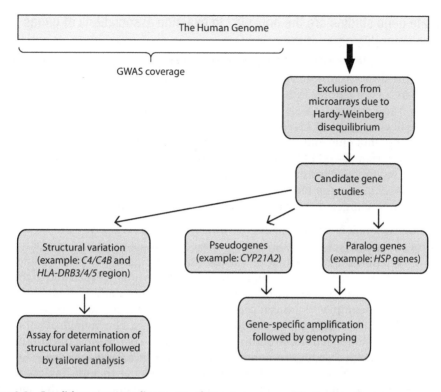

Figure 8.3. Candidate gene studies to complement genome-wide studies. Paralog genes and pseudogenes are gene copies with high sequence similarity to the target genes. They cannot be genotyped by microarrays due to confounding of the results by co-amplification of paralogs or pseudogenes together with the target genes. This issue is identified by Hardy–Weinberg disequilibrium and these SNPs fail at the quality control step of microarray production and are excluded. They can still be genotyped in candidate gene studies: target genes can be selectively amplified by the careful design of PCR assays to exclude paralogs and pseudogenes from amplification and then using the PCR products for genotyping. For structural variation, the first assay can determine which variant (for example *HLA-DRB3* or *-DRB4* or *-DRB5*) is present in each sample. Then, the analysis can be restricted to the SNPs that exist in the structural variant(s) present in each individual.

PCR can genotyping be done with no confounding by the paralogs. Heat shock protein genes are examples of a large family of paralogs. The *HSPA1B* variant rs1061581 (better known as the 1267 A > G *Pst*I polymorphism) has shown many associations in candidate gene studies but has never been assessed in a GWAS due to difficulties with including it in a genotyping chip. Likewise, the gene encoding 21-hydroxylase, *CYP21A2*, which is involved in adrenal steroid synthesis, remains the only CYP gene that has never been assessed in a GWAS for breast cancer, as it has its pseudogene next to it. The pseudogene has more than 99% sequence similarity to the expressed gene and contains all of the mutations that cause the most common Mendelian disorder of childhood (congenital adrenal hyperplasia) when they occur in the expressed gene. In clinical genetics laboratories, *CYP21A2* mutations are routinely genotyped using low-throughput manual methods, but cannot be genotyped using GWAS chips. An example is rs6471 (V282L) that causes late-onset adrenal hyperplasia characterized by hyperandrogenism for which no GWAS result can be found.

Candidate gene studies are the only way to examine associations of genes in highly polymorphic parts of the genome

Since SNP genotyping in GWAS chips uses sequences that are very close to the SNP, SNPs flanked by highly polymorphic sequences are very hard to include. One of the most polymorphic parts of the genome is the HLA region. The *HLA-DQA1* 3′ UTR is extremely polymorphic and genotyping these SNPs is very difficult by any method other than sequencing. At the time of writing, NCBI ENTREZ SNP lists more than 100 SNPs in the *HLA-DQA1* 3′ UTR (<1000 bp), but only four of them have HapMap data. Two SNPs in this region have particular importance—rs1142316, known as the *Bgl*II polymorphism, and rs9272863, identified bioinformatically as a slice-site regulatory SNP—but these will never be examined in a GWAS due to the technical difficulties caused by the extremely polymorphic flanking regions. A targeted approach is needed to obtain information on the gene.

For the HLA region, it is not only sequence but also variation in gene content, known as **structural variation** (see Figure 8.3), that poses difficulties to microarray genotyping. Structural variation refers to situations where the same genomic position may have alternative genomic content in different individuals that is not incorporated in the reference sequence. As there is as yet no sequence for the cumulative genome (the pangenome), only genes and variations represented in the reference sequence are currently being examined. The HLA complex provides a good example. The biological importance of the HLA-DR region makes its structural variations some of the most important in the genome. The reference sequence contains *HLA-DRB5* in addition to the *HLA-DRB1* gene in a structural variation region, but does not acknowledge that the haplotypes containing this gene only make up 15% of all haplotypes in the population. The remaining chromosomes in the population contain either *HLA-DRB3* or *HLA-DRB4* in the same position, or no gene at all. The 1000 Genomes Browser (URL given at the end of the chapter) lists polymorphisms for *HLA-DRB5* but none for *HLA-DRB4*. However, *HLA-DRB4* and *HLA-DRB3* have polymorphisms that are routinely typed in tissue typing laboratories but which are not included in GWAS chips because they are not listed in reference sequences. If they were included in chip genotyping, they would be excluded during quality control steps, as the missing call rates would be very high simply because a lot of chromosomes do not carry them at all.

These examples justify the ongoing need for candidate gene studies, and particularly sequencing efforts directed at candidate genes or genomic regions. The SNPs that cannot be included in GWAS chips may be represented by others that correlate with missing SNPs, but this possibility has not been systematically examined. Results obtained in GWASs are valid, but there may be results that remain to be obtained, and the exclusive use of high-throughput methods may leave more work to do by other methods.

8.4 Advantages and Disadvantages of a GWAS

The GWAS is a powerful study design that has many advantages, but there are still potential deficiencies that must be considered. Candidate gene studies also have major drawbacks.

Candidate gene studies are plagued by irreproducibility

Although the very first genetic association between blood groups and stomach cancer has stood the test of time and recent GWASs have identified these associations, most candidate gene studies in general yield inconsistent results, even in other cancers. One reason for inconsistency is that candidate gene studies are typically small, exploratory studies that suffer from results due to chance. One survey of more than 600 positive candidate gene study associations showed that 166 of the loci were studied three or more times, but only six could be consistently replicated. While this looks poor, there may be genuine reasons for replication failure, particularly in different populations where the interacting environmental exposures may be different, and genetic heterogeneity and variable gene-gene interactions may generate different results. Overall, candidate gene studies are considered to have had a ratio of false positive to false negative of more than 100:1, as opposed to lower than 1:100 in GWASs. This huge difference is likely to result from differences in the stringency levels in the design, execution, and analysis of these two types of studies. Another survey examined past candidate gene study results in subsequent GWASs for the same phenotypes. Few of the numerous candidate gene associations could be verified in a GWAS, but those that were verified had large effect sizes. These results highlight the fact that most of the past candidate gene studies were only powered to detect large effect sizes and, by inference, most of the associations with small and modest effect sizes were chance findings (false positives) in underpowered studies.

By following the same stringent conditions applied to a GWAS, such as very large sample size, a built-in second study for replication, and strict statistical safeguards, a much higher proportion of candidate gene studies could be replicated successfully. Candidate gene studies must also incorporate ancestry informative markers (AIMs) to minimize the potential for population structure to invalidate results.

GWASs have several advantages over candidate gene studies

Most researchers are turning to GWAS platforms even for smaller pilot studies and GWASs have become the gold standard for association studies with common complex disorders. However, a GWAS is characterized by more than the use of genome-wide genotyping. A typical GWAS has important differences from a typical candidate gene study (**Table 8.1**). An appropriately designed GWAS produces a true positive result that is already replicated, and only leaves identification of the causal variant to future studies (which is the topic of Chapter 10).

GWAS design inherently takes care of the major threat to the validity of any genetic association study—population stratification—by including thousands of AIMs for statistical adjustment of the results. Another advantage of a GWAS is the inclusion of both sequence variants and copy number variation assessment in the design.

GWAS workflows are strictly designed

A typical GWAS has a very strictly designed workflow, partly reflecting the lessons learned from earlier genetic association studies and partly addressing the issues unique to a GWAS (**Figure 8.4**). It is because of this strict protocol that "top-hit" associations in a GWAS are considered as true associations rather than potential chance findings. Equally importantly, a negative result in a GWAS almost certainly rules out a major genetic component in disease etiology.

GWASs have had some spectacular results

The GWASs conducted to date have unraveled unexpected associations that would have otherwise remained undetected. The most notable and illustrative example is the demonstration of associations between several SNPs in chromosome 8q24 and multiple cancers.

Table 8.1 Main characteristics of candidate gene studies and genome-wide association studies

Features	Candidate gene study	GWAS
Hypothesis-driven	Yes (based on linkage results or biology)	Hypothesis-free (agnostic)
Sample size	Generally relatively small (hundreds)	Generally large (thousands)
Coverage	Selected gene(s)	Whole genome
SNPs	Fewer than 1000	Up to 5 million
CNVs	May be included	Included in most
STRs	May be included	Not included
Genotyping	Low-throughput methods or custom microarray	Microarray (GWAS chip)
Quality control tests	Basic	Extensive
Analysis	Less stringent	Very stringent (uses genome-wide statistical significance threshold)
Adjustment for ancestry	Not routinely done	Always incorporated
False-positivity rate	Very high	Very low
False-negativity rate	Low	Potentially high
Replication	Required, but usually in a later study	Always incorporated

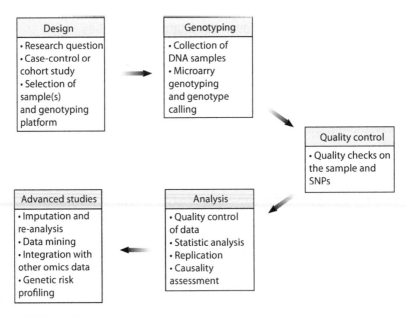

Figure 8.4. GWAS workflow.

These SNPs are in the middle of a very large gene desert and no candidate gene study would ever have examined them. Likewise, cell cycle regulator gene SNPs have shown associations in coronary artery disease and type 2 diabetes, which would not have been considered in candidate gene studies. Another physiological pathway that unexpectedly turned out to be involved in disease biology is autophagy in Crohn disease. Some lessons we have learned from GWASs are:

- Even gene deserts, which would not have been explored in candidate gene studies, may have functional variants.

- Most effect sizes in associations of common SNPs with complex disease susceptibility are small (OR ≤ 1.50), but for rare alleles the median OR goes up to 2.2.

- Associations with small effect sizes may still lead to successful clinical translations, especially for pharmacologic interventions.

- Most associations (>90%) are with SNPs that are outside coding regions and associations with them are mainly due to their regulatory effects on gene expression.

- The gene nearest to a GWAS association signal is not necessarily the causal gene, and the top SNP may be an eQTL for a different gene that may even reside in a different chromosome.

- Only a small proportion of the previous genetic associations reported in the candidate gene study era has been replicated in GWASs. However, not all previous observations can be re-examined by the GWAS design (due to coverage limitations).

- Even after multiple GWASs have been performed in multiple common diseases, the proportion of heritability explained remains small. One implication of this observation is that DNA sequence variants are unable to explain all heritability.

- Genetic associations are more useful in probing disease biology rather than developing predictive models for disease development.

GWASs do not usually find variants with large effect sizes

When the GWAS era began, it was a very promising beginning. The first study published in 2005 used a chip with 10,000 SNPs and found an unexpected association in a sample size of 100 with an impressive effect size (OR > 4.0). That finding in age-related macular degeneration has stood the test of time, but no study since then has come up with a result equally as impressive. The current standard for the number of SNPs included in GWAS chips is 5 million, and even more can be imputed, sometimes in more than 100,000 individuals. The P values continue to get smaller, especially when multiple studies are combined (as in a **meta-analysis**), but large effect sizes are rare. Candidate gene study results that, rarely, can be replicated in a GWAS usually yield effect sizes larger than average for a GWAS. The effect sizes measured by the OR in SNP associations identified in a GWAS are usually less than 1.5. However, an analysis of GWAS associations with SNPs that are uncommon (minor allele frequency is between 0.01 and 0.05) showed that the median OR was 2.2 (interquartile range 1.4 to 3.4). Thus, it is possible to have a large OR for an individual SNP association, but common variants typically yield small to modest ORs. Such low ORs have hampered efforts to generate genetic risk profiles that may replace or at least improve existing risk profiles for common diseases such as breast cancer or type 2 diabetes. While this has been a disappointment, GWAS results have contributed enormously to the understanding of disease biology, regardless of the magnitude of the effect sizes. Low-risk alleles have been of little predictive value but are very informative in determining pathogenic biochemical pathways.

The GWAS has not solved the problem of missing heritability

In the beginning, GWASs were thought to be the ultimate tool to uncover a large proportion of heritability. This expectation has not been realized and we now face a **missing heritability** problem. Most GWASs performed to date have been based on chips that contained common SNPs due to the common disease–common variant theory. The end result is that GWASs appear to have only identified common alleles with small effects. It has, however, been shown that untyped rare variants can create associations that are credited to typed common variants. With increasing recognition of the role played by rare variants in disease pathogenesis, most GWAS results are likely to be refined in follow-up studies, mainly by sequencing of the regions that revealed the strongest disease association signals. Missing heritability is a more complex issue to solve than can be solved by typing rare variants, and **epigenetic variation** is very likely to play a big role.

GWAS design means false-negative results are common

The most unexpected and potentially harmful disappointment of the GWAS design has been the rate of false-negative results. GWASs are designed to lower the false-positive result rates, mainly by implementing a very strict genome-wide statistical significance threshold. Before the GWAS era, common diseases were thought to be caused by several hundred SNPs acting in concert; however, in no disease have GWASs found more than a few dozen SNPs cumulatively associated with disease risk. It is clear that some associations have been missed due to the strict statistical threshold. Given that statistical power is not uniformly distributed across a range of SNPs with different frequencies and different effect sizes for their associations with the phenotype, using a single threshold for all of them inevitably causes some false negatives. It has been pointed out that the most extreme P value alone is an inadequate predictor of true effects and the P value does not say much about biological significance. An example is the *PPARG* SNP rs1801282, which was known from candidate gene studies to be associated with type 2 diabetes risk and has well-supported biological plausibility. In the first three GWASs in type 2 diabetes, this SNP did not show an association with statistical significance exceeding the genome-wide threshold. The first three large studies (WTCCC, FUSION, and DGI) had to be pooled for this SNP to show as a GWAS result.

The historical epidemiologic approach has been to minimize type II errors (false negatives), but statisticians are more wary about minimizing type I errors (false positives). In GWASs, it is the latter statistical approach that has been adopted. The failure to test for different genetic models, joint effects, interactions, effect modifications such as sex-specific associations, or haplotype associations mainly resulted from trying to avoid increasing the number of comparisons that would lead to even more statistical adjustments. These omissions may have contributed to the missing heritability problem. As in the *PPARG* example given above, we now rely on meta-analysis of pooled data to find associations reaching the statistical threshold adopted for GWASs. The evidence for the presence of false negatives comes from studies that have shown that if the risk profiling is not based on just the few top hits but uses a larger number of top SNPs, or in fact the whole set of SNPs, the results become more promising. Although an extensive amount of research is ongoing in this field, we are currently missing the ideal statistical or data-mining methodology for the best use of GWAS data.

More powerful techniques are needed to identify causal variants

The main function of a GWAS is to discover the location of the disease gene rather than to characterize the causal variant. Thus, even in a GWAS, an associated variant is not necessarily a true causal variant. Causal variant detection may need even more advanced approaches such as whole-genome sequencing using next-generation sequencing. The statistical correlations found in association studies must be confirmed by functional studies—real, biological experiments are still needed, which can then lead to translation to medical uses. The post-genomic era is expected to make use of the existing GWAS data to achieve the ultimate aim of genetic association studies by allowing translation of the findings to practical applications.

Key Points

- Candidate gene studies still have value in genetic epidemiology as long as the same stringent conditions of a GWAS are applied to them, plus they incorporate markers that cannot be examined in GWAS platforms.

- Variant selection is an issue for both candidate gene studies and GWASs, and it serves different purposes for each.

- TagSNPs help avoid redundancy but may not represent all polymorphisms within a gene or region. Other SNPs that cannot be tagged may need to be included in the genotyping scheme for complete coverage.

- The causal SNP is rarely the top hit, and is not necessarily in the gene nearest to the top hit. The top hit may be an eQTL for a gene that resides in a different chromosome.

- GWASs have achieved a lot but should not be seen as the ultimate tool for dissecting the genetic basis of disease susceptibility.

- Current GWAS results are no doubt valid, but a number of associations may have been missed.

- Even when all DNA sequence variant associations are identified, a portion of heritability will remain unaccounted for (missing heritability) due to epistatic and epigenetic factors.

URL List

Variants in a gene

1000 Genomes Browser. National Center for Biotechnology Information. http://www.ncbi.nlm.nih.gov/variation/tools/1000genomes

Ensembl. Wellcome Trust Sanger Institute/EMBL-EBI. http://www.ensembl.org

ENTREZ SNP (dbSNP). National Center for Biotechnology Information. http://www.ncbi.nlm.nih.gov/snp

Genome Bioinformatics. University of California Santa Cruz. http://genome.cse.ucsc.edu

International HapMap Project. http://hapmap.ncbi.nlm.nih.gov

Already known genetic associations

A Catalog of Published Genome-Wide Association Studies. National Human Genome Research Institute. National Institutes of Health. http://www.genome.gov/gwastudies

dbGaP. Association results browser. National Center for Biotechnology Information. http://www.ncbi.nlm.nih.gov/projects/gapplusprev/sgap_plus.htm

GAD (Genetic Associations Database). National Institutes of Health. http://geneticassociationdb.nih.gov

GRASP (Genome-Wide Repository of Associations Between SNPs and Phenotypes). National Heart, Lung, and Blood Institute. National Institutes of Health. http://apps.nhlbi.nih.gov/Grasp

GWAS Central. http://www.gwascentral.org

HuGE Navigator. http://hugenavigator.net/HuGENavigator/home.do

PubMed. National Center for Biotechnology Information. http://www.ncbi.nlm.nih.gov/pubmed

Bioinformatics tools

eQTL Browswer (UChicago). http://eqtl.
uchicago.edu/cgi-bin/gbrowse/eqtl

Regulome DB. Center for Genomics and
Personalized Medicine at Stanford University.
http://regulomedb.org

SCAN (SNP and CNV Annotation Database).
University of Chicago. http://www.scandb.org

Bioinformatics suites

F-SNP. Queen's University. http://compbio.
cs.queensu.ca/F-SNP

GenEpi Toolbox. http://genepi_toolbox.i-med.
ac.at/?page_id=329

GWASdb. http://jjwanglab.org:8080/gwasdb

PheGenI. Phenotype-Genotype
Integrator. National Center for Biotechnology
Information. http://www.ncbi.nlm.nih.gov/
gap/phegeni

SNPnexus. Barts Cancer Institute. http://www.
snp-nexus.org

Further Reading

Gonzaga-Jauregui C, Lupski JR & Gibbs RA
(2012) Human genome sequencing in health
and disease. *Annu Rev Med* 63, 35–61 (doi:
10.1146/annurev-med-051010-162644).
(*This paper reviews the advances made in
understanding the whole genome and the
relevance of sequencing to the identification of
disease-associated variants.*)

Hattersley AT & McCarthy MI (2005)
What makes a good genetic association
study? *Lancet* 366, 1315–1323 (doi: 10.1016/
S0140-6736(05)67531-9). (*This review discusses
issues most relevant to genetic association
studies as part of a series in The Lancet.*)

Healy DG (2006) Case-control studies in the
genomic era: a clinician's guide. *Lancet Neurol*
5, 701–707 (doi: 10.1016/S1474-4422(06)70524-
5). (*A review on the most common epidemiologic
study design in genetic association studies, and
its main problem of nonreplication.*)

Hirschhorn JN & Daly MJ (2005) Genome-wide
association studies for common diseases and
complex traits. *Nat Rev Genet* 6, 95–108 (doi:
10.1038/nrg1521). (*One of the seminal papers
that set the scene for forthcoming genome-wide
association studies.*)

Iles MM (2011) Genome-wide association
studies. *Methods Mol Biol* 713, 89–103 (doi:
10.1007/978-1-60327-416-6_7).

Manolio TA & Collins FS (2009) The
HapMap and genome-wide association
studies in diagnosis and therapy. *Annu Rev
Med* 60, 443–456 (doi: 10.1146/annurev.
med.60.061907.093117). (*Covers the genome-
wide association study—its triumphs as well as
its limitations.*)

Tabor HK, Risch NJ & Myers RM
(2002) Candidate-gene approaches for
studying complex genetic traits: practical
considerations. *Nat Rev Genet* 3, 391–397
(doi:10.1038/nrg796). (*A concise practical guide
to candidate gene studies. SNP selection is well
covered. It also provides a balanced critique of
candidate gene studies.*)

Teare MD (2011) Candidate gene association
studies. *Methods Mol Biol* 713, 105–117 (doi:
10.1007/978-1-60327-416-6_8).

Pettersson FH, Anderson CA, Clarke GM
et al. (2009) Marker selection for genetic
case-control association studies. *Nat Protoc* 4,
743–752 (doi: 10.1038/nprot.2009.38).

Williams SM, Canter JA, Crawford DC
et al. (2007) Problems with genome-wide
association studies. *Science* 316, 1840–1842
(doi: 10.1126/science.316.5833.1840c). (*A
commentary written in the early years of the
GWAS era to warn about the view that obvious
success should not lead to the conclusion that
"GWAS is a panacea."*)

Zintzaras E & Lau J (2008) Synthesis of genetic
association studies for pertinent gene-
disease associations requires appropriate
methodological and statistical approaches.
J Clin Epidemiol 61, 634–645 (doi: 10.1016/j.
jclinepi.2007.12.011). (*A useful critique of
genetic association studies from epidemiologic
and statistical points of view. Also provides an
insight into meta-analysis.*)

Statistical Analysis of Genetic Association Study Results

9

Genetic association studies examine relationships between the presence of a genotype and a phenotype. Under this general aim, there are several nuances: the genotype may in fact be several genotypes acting jointly; the relationship may only be applicable to a subset of the disease (especially in heterogeneous diseases such as asthma or depression); or the relationship may be apparent only in the presence of another factor. Correlation does not mean causation, and as the aim is to find causality, some assessment of potential causality should ideally be carried out. In this chapter, standard association analysis will be reviewed, followed by discussion of more advanced approaches to association analysis. The latter half of the chapter is devoted to GWAS analysis, which has unique features and quality control steps.

9.1 Standard Genetic Association Analysis

An association is sought between smoking and lung cancer, for example, by examining whether there are more smokers (smoking-positive group) among lung cancer cases (lung cancer-positive) compared with healthy controls. In environmental epidemiology, the presence or absence of the environmental exposure is the main explanatory variable. In any genetic association study, alleles, genotypes, and haplotypes are the main variables, and their examination requires a different approach. A genetic association is most commonly analyzed by considering genotypes. Since there are three genotypes for each SNP (AA, AB, and BB), their analysis requires some manipulation. For analysis, the three genotypes are coded according to the genetic model of interest (see Chapter 4).

Dominant, recessive, and overdominant models are analyzed in the same way

The dominant, recessive, and overdominant models require the three genotypes to be collapsed into two by treating the two variant allele-positive genotypes as equally relevant (**Figure 9.1**). In order to determine whether the dominant model fits the data, the heterozygote (AB) and variant homozygote (BB) genotypes are coded as 1 and the wild-type genotype (AA) is coded as 0. In the case of the recessive model, the assumption is that the variant allele is associated with risk when present in two copies. To do the analysis for the recessive model, BB is coded as 1 and other genotypes as 0. In the heterozygote advantage or overdominant model, the interest is in heterozygosity as opposed to the two homozygote genotypes. The AB genotype is coded as 1 and homozygote genotypes (AA and BB) as 0. The overdominant model is rarely used but is appropriate in studies of genomic regions where heterozygote advantage has already been observed, such as the HLA region. The only difference between these models is in which genotype or collapsed genotypes they consider as the risk genotype: the dominant model or variant allele-positivity considers AB and BB as risk genotypes; the recessive model considers BB as the risk genotype. As discussed in Chapter 6, the statistical properties of the association are then assessed by

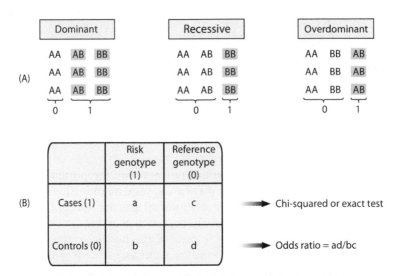

Figure 9.1 Coding and 2 × 2 table analysis of dominant, recessive, and overdominant models.
In each of the dominant, recessive, and overdominant models, the reference genotypes(s) are coded as 0 and risk genotype(s) are coded as 1. By converting the three genotypes to a binary variable, coded as 0 or 1, the data can be analyzed by constructing 2 × 2 tables and entering the counts of reference genotypes (0) and risk genotypes (1) for cases (1) and controls (0) in each of the four cells.

constructing a simple 2 × 2 table and applying either the **Chi-squared test** or **Fisher's exact test**. Only if a statistical adjustment is needed, for example for age or sex, are advanced 2 × 2 table-based methods or logistic regression used.

Additive model analysis uses all three genotypes

The additive model examines the gene-dosage effect by using all three genotypes and taking into account the gradual change, also known as the trend in risk, associated with each genotype from AA (referent) to AB to BB. Since the genotype AB has one copy of the variant allele B, and the genotype BB has two copies, a gradual change in the risk associated with these genotypes in the same direction would suggest a gene-dosage effect. In additive model analysis, no collapsing of the genotypes is involved in the coding of genotypes: the genotypes AA, AB, and BB are coded as 0, 1, and 2 (although the codes can be any three numbers, as long as they are ordered from smaller to larger, such as −1, 0, and 1) reflecting the number of variant alleles in each genotype (**Figure 9.2**). This type of analysis has greater statistical power; in other words, the ability to detect statistical significance will be greater in a trend analysis compared with the analysis of the same data after collapsing the three genotypes into two and analyzing them by other models. The crucial point is that the additive model analysis will detect an association if the effect sizes change in a linear or near-linear manner. The statistical test examines whether the risk conferred by AB and BB changes in the same direction, and how closely the change resembles a straight line. As shown in **Figure 9.3A**, the additive model fits the result when the change in effect size (odds ratio) is linear. The more the deviance from the straight line, the less statistically significant is the fit to the additive model. If the change is not linear at all, additive model analysis will not reveal any statistical significance. In real life, the additive model is the most biologically plausible model, even though the change may not be perfectly linear.

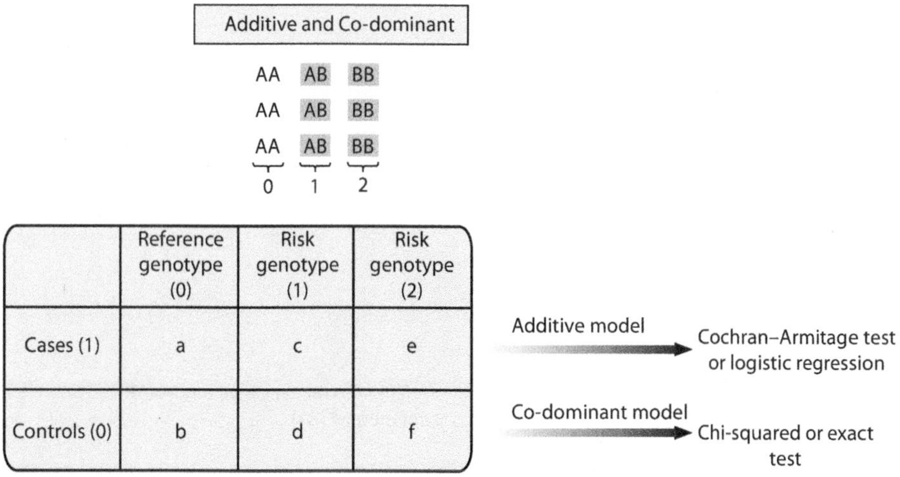

Figure 9.2 Coding and 2 × 3 table analysis of additive and co-dominant models. No genotype groups are collapsed and all three genotypes are coded. The resulting 2 × 3 table is analyzed differently for the additive and co-dominant models.

The additive model is currently the first-choice genetic model in association analysis for several reasons. It examines the gene-dosage effect (one of the causality assessment criteria), it is the only model that can be used even if HWE cannot be confirmed, and quantitative genetics has shown that it is the most relevant model for complex phenotypes. Analysis using the additive model is generally sufficiently powerful to detect associations that do not fit into the gene-dosage effect, such as the dominant and recessive models shown in **Figure 9.3B and C**, but important exceptions exist. Associations fitting to strictly dominant or recessive models may not be detected by the exclusive testing for the additive model. The overdominant model (**Figure 9.3D**) never yields a linear change in risk estimates and would always be missed by the additive model tests. Simulation studies and empirical data strongly suggest that associations fitting the recessive model with the causal variant allele frequency not close to 0.50 are the hardest to detect by additive model analysis. Likewise, in an indirect association, if LD is imperfect, even if the association of the causal allele would have fitted the additive model, the additive model will be distorted as a function of the strength of the LD (as measured by r^2). Thus, in practice, the additive model is assumed to fit the data, but it may not. Examination of additional models is usually avoided out of concerns for multiple comparisons, but if avoiding false negatives has high priority and safeguards are in place to rule out false positives, all models should be considered in a thorough analysis of genetic association data.

Additive model analysis requires either the use of the **Cochran–Armitage trend test** (a 2 × 3 table analysis) or a regression model, which would be logistic regression if the outcome is binary (two possible outcomes, such as case or control status in a case-control study). In a 2 × 3 table, all three genotypes are considered, and it is possible to get an odds ratio for the risk conferred by the AB genotype and, separately, the BB genotype (using the genotype AA as the baseline or referent). More commonly, however, a single odds ratio is presented for this analysis. This is the average change in risk for each variant allele in the genotype. This averaged single odds ratio is called the per allele, allelic, or **common odds ratio**. In contemporary genetic association analysis, unless stated otherwise, the default analysis uses the additive model and produces the common odds ratio together with its

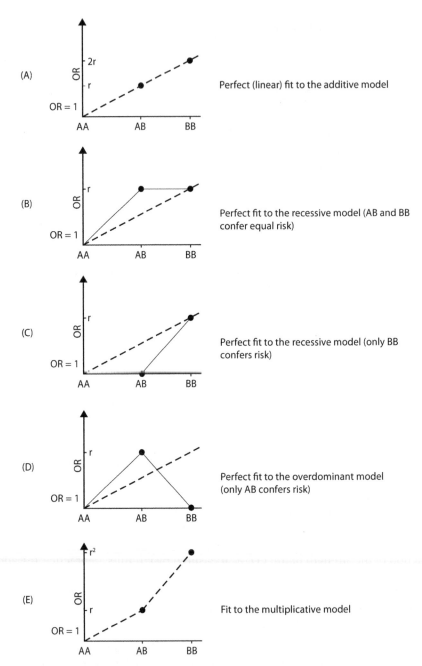

(A) Perfect (linear) fit to the additive model

(B) Perfect fit to the recessive model (AB and BB confer equal risk)

(C) Perfect fit to the recessive model (only BB confers risk)

(D) Perfect fit to the overdominant model (only AB confers risk)

(E) Fit to the multiplicative model

Figure 9.3 The change in the effect size for different genetic models and the correlation of the pattern of change to genetic risk. The dashed line in each graph represents the perfect linear trend in the risk increase; a dotted line represents the actual risk change for each scenario. By using it as the referent, the odds ratio (OR) for genotype AA is fixed at 1.0. (A) The additive model fits perfectly when the change in the OR is linear. When the change in OR is not linear, the additive model does not fit perfectly and the shape of the graph indicates the correct model: (B) the dominant model and (C) the recessive model show an imperfect linear fit; and (D) the overdominant model shows no linearity. (E) The multiplicative model uses allele counts and gives a perfect fit when the OR for the homozygous genotype is the square of the OR for the heterozygous genotype.

confidence interval and a *P* value for its statistical significance. There are online calculators that can compute the *P* value and the common odds ratio for the additive model using the genotype counts in controls and cases (for example, the DeFinetti program; see the URL list at the end of the chapter).

Box 9.1 contains worked examples of a simple genetic association study analysis for the different genetic models described in this chapter and shows how using different models

---

Box 9.1 Statistical analysis of a simple case-control study for different genetic models*

In a case-control study, the genotype counts were as follows**:

	AA	AB	BB
Cases (*n* = 142)	28	92	22
Controls (*n* = 161)	58	71	32

Analysis for each genetic model is carried out as follows:

DOMINANT	(Risk)	(Reference)
	AB + BB	AA
Cases (1)	114	28
Controls (0)	103	58

Odds ratio = (114 × 58) / (103 × 28) = 2.29; *P* = 0.002 (χ^2 = 9.87).

RECESSIVE	(Risk)	(Reference)
	BB	AA + AB
Cases (1)	22	120
Controls (0)	32	129

Odds ratio = (22 × 129) / (32 × 120) = 0.74; *P* = 0.32 (χ^2 = 0.99).

OVERDOMINANT	(Risk)	(Reference)
	AB	AA + BB
Cases (1)	92	50
Controls (0)	71	90

Odds ratio = (92 × 90) / (71 × 50) = 2.33; *P* = 0.003 (χ^2 = 12.99).

ADDITIVE***	(Risk)	(Risk)	(Reference)
	BB	AB	AA
Cases (1)	22	92	28
Controls (0)	32	71	58

Additive model: Common odds ratio = 1.25; *P* = 0.12 (χ^2 = 2.38; Cochran–Armitage trend test).

Co-dominant model: *P* = 0.001 (χ^2 = 13.89).

*The Further Reading section lists a full protocol for complete genetic association analysis using PLINK on the data sets provided; **these genotype counts are from the DeFinetti online calculator demonstration. Results for dominant, recessive, and additive models can be compared with the results of the demonstration; ***the co-dominant model uses the same 2 × 3 table.

can yield very different results. The highly significant result for the dominant model (AB + BB) is actually a reflection of the overdominant association (AB), since the recessive model (BB) does not show an association. Because of the nonlinearity of the risk change, the additive model shows no significant result, but the co-dominant model strongly picks up the deviation in genotype counts between cases and controls.

Co-dominant model analysis also uses all three genotypes

The **co-dominant model**, also called the **general model** or unrestricted model, uses all three genotype counts, but does not assume a linear change in risk. It is appropriate to use if the risk conferred by the genotype AB lies between that of AA and BB, but not necessarily near the straight line. Co-dominant model analysis uses the three genotypes as separate categories in a 2×3 table (see Figure 9.2) and any change in any direction between the genotype counts of cases and controls is assessed. In statistical terms, the test examines the independence of the values in the 2×3 table. The co-dominant model is a good starting point when the genetic model is unknown because it does not impose a specific model; it represents a model-free approach to association analysis and therefore avoids the danger of obtaining misleading estimates of risk from analyses based on inappropriate model selection. Although the co-dominant model will detect any change in genotype counts between cases and controls fitting to any of the models shown in Figure 9.3, it is not as powerful as the correct genetic model that represents the data better. As a result, the P value will be smaller when the correct genetic model is tested. Like a 2×2 table, a 2×3 table can also be analyzed by the Chi-squared test for the co-dominant model. The additive model testing, however, requires some computing power.

Allele counts are compared to test the multiplicative model

The statistical models presented above are genotype based, meaning that genotype counts are used in the assessment. The **multiplicative genetic risk model**, also called the **allelic model**, uses allele counts (frequencies). While the additive model is the best-fitting genetic model when the odds ratio for homozygosity is twice as large (2r) as the odds ratio for the heterozygous genotype (r), the multiplicative model is the best fit when the risk increases multiplicatively and the odds ratio for the homozygous genotype is r^2 (**Figure 9.3E**). The multiplicative model uses allele frequencies (for allele A and allele B) derived from the genotype (AA, AB, and BB) counts for cases and controls. This model is not used frequently for two reasons. First, the allele counts are derived from genotype counts, and this requires genotype counts conforming to HWE. The model assumes that the alleles A and B in each genotype are independent and show no association with the phenotype. Second, the unit of analysis is not the person but the chromosomes, because genotypes are counted in persons while alleles are counted in chromosomes, and an odds ratio for the risk for each chromosome is hard to interpret. However, the multiplicative model is useful for initial screening of the data to see whether there is any signal worth pursuing for examination by other genetic risk models. The use of chromosomes as the unit of analysis doubles the sample size and increases the statistical power to detect differences that may be missed by other genetic models.

Haplotype analysis can identify associations due to unknown rare variants

Haplotypes are linear arrangements of nucleotides residing on the same chromosome (see Chapter 1). The higher the number of nucleotides forming a haplotype, the higher is the number of possible haplotypes, some of which will occur at low frequency. Rare

	SNP1	SNP2	SNP3	Unknown SNP	SNP4	SNP5	SNP6
Haplotype 1:	T	C	T	C	C	A	G
Haplotype 2:	C	T	T	C	T	T	G
Haplotype 3:	C	T	G	C	T	T	G
Haplotype 4:	C	_C_	_G_	_T_	_T_	_A_	A
Haplotype 5:	C	T	T	C	T	A	G
Haplotype 6:	C	T	G	C	T	T	G
Haplotype 7:	T	C	G	C	C	A	A

Rare variant C > T

Figure 9.4 Haplotype analysis may unravel associations with untyped, unknown rare alleles.
The seven haplotypes shown consist of six known SNPs and an as-yet unknown SNP in the middle of the haplotype (shown in a dotted box). Its allele, T, is rare and is on a unique haplotype. The two nucleotides (C and G) to the left of the rare allele T, and the two nucleotides (T and A) to the right of it, are only present in haplotype 4, which also carries the rare allele T. The allele T is unknown, but the detection of the haplotype C-G-T-A (underlined) would indicate the presence of the unknown allele T, and this haplotype would show an association if the rare allele T is a causal allele, even if none of the variants flanking it would show an individual association.

haplotypes that are made up of common and genotyped alleles may harbor a causal rare allele, which is unknown and not yet typed (**Figure 9.4**). Haplotype analysis can therefore reveal a strong association, even in a study that genotyped only common alleles, and even if none of the common variants have yielded an individual association. Such an association is usually attributed to a rare allele that was not genotyped but was present in one of the haplotypes in-between the genotyped alleles. In the past, haplotyping was only possible when family data were available, but it can now be achieved by using computational algorithms (such as PHASE or UNPHASED) without needing family data. Thus, there is no excuse to not use this approach if common variants have not revealed an association. Even if a positive result has already been found, haplotype analysis can be done to examine if the association will get stronger, and if the causal variant can be localized with greater precision. Although a stronger haplotype association, compared with common allelic associations, may not reveal the rare causal allele, it will pinpoint the haplotype that harbors the causal allele. Haplotypes may also show stronger associations than individual variants if two markers are on the same haplotype and their joint effect is stronger than the sum of their individual effects.

Logistic regression has advantages

Logistic regression is the statistical method of choice for analysis of an association study because it allows inclusion of potential **confounders** or **covariates** into the computation of the adjusted odds ratio and statistical significance. A similar adjustment can be achieved for one or two confounders by constructing different 2×2 tables and then averaging the results by taking their weight into account, which is known as the **Mantel–Haenszel procedure**. By using logistic regression, one can add additional variables to the analysis as necessary, although there is a limit to the number of additional variables that should be included. A good rule of thumb is a maximum of one variable per 10 cases. For example, in a small study including only 50 cases, at most five variables can be included in the adjusted analysis. It has to be remembered that the minimum possible use of covariates is desired in statistical analysis. Another rule of thumb is not to have more covariates than the number of data points. Trying to create a complex predictive model with lots of covariates may create an **overfitting** problem. It may prove to be even more difficult to replicate an overfitted or heavily adjusted statistical model. It is best to have a very large sample size and decrease the number of potential confounders to a small number using established statistical methods.

When a logistic regression analysis is done with adjustments for multiple potential confounders, it is commonly called a **multivariate analysis**. What is actually meant is a multivariable analysis. Multivariate analysis is a very special and advanced statistical method and the use of this term is best reserved to the occasions when it means what it is.

Genetic associations can be confounded

In epidemiologic studies, any observed association may be confounded and any possible confounding should be ruled out. A confounded association is an association that is not with a causal factor but rather with another factor correlated with the causal factor. If a confounded association is considered as a causal one by mistake, any resulting intervention, like drug development, will not be useful because it will not be targeted to a causal factor. In a genetic association study, the most common genetic cause of confounding is the widespread linkage disequilibrium (LD) that results in multiple associations, one of which may or may not be with the causal variant. Only a direct association is a causal one and indirect associations are results of confounding by locus (Chapter 4). Almost every variant is in LD with a few other variants. A survey of GWAS associations in autoimmune disorders revealed that only 5% of the GWAS associations are likely to be causal (direct); the others are indirect associations. In a genetic association study, confounding by locus may be reduced by including as many variants as possible in the study, but the causal one may be an as-yet unknown rare variant. Ultimately, only sequencing studies will identify the causal variant with confidence, as has been observed in Mendelian or single-gene disorders.

Confounding by nongenetic factors is also possible in a genetic association study. The most critical potential confounder is population substructure (also known as population stratification) or confounding by ethnicity (Chapter 4). It is important that this possibility is considered in each study. The minimum effort to address population stratification includes matching cases and controls for geography, ancestry, and ethnicity, or collecting data on these variables, especially self-declared ethnicity, for statistical adjustment. As shown in **Figure 9.5**, differences in the frequencies of ethnic groups in cases and controls may create a spurious association (due to confounding by ethnicity) that disappears when ethnicity-matching is applied to the analysis. In essence, this is what statistical adjustment does: the association statistics for each of the variable categories (such as Europeans and non-Europeans in the sample) are estimated and a weighted average of the statistics is provided. The usual potential confounders such as age, gender, and socioeconomic levels are also relevant, but not to the same extent as in nongenetic studies. This is because genotypes are not expected to segregate with these potential confounders (with the exception of sex chromosome variants between males and females). Since genotypes do not change in frequency between socioeconomic groups or age groups (except maybe in the extreme age groups), genetic association studies are less likely to be confounded by these factors.

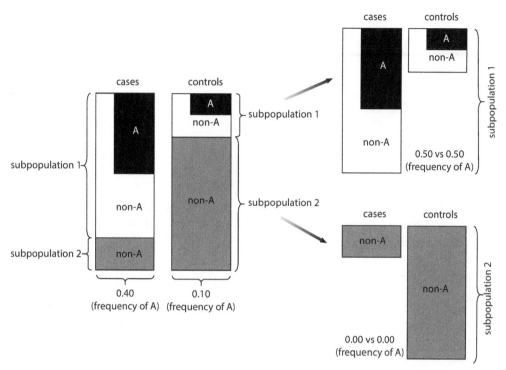

Figure 9.5 Confounding by ethnicity in a disease association study with blood group A.
The sample consists of two subpopulations: subpopulation 1 makes up 80% of cases (because
the disease is more common in subpopulation 1) and has 50% frequency for blood group A.
Subpopulation 2 has no individuals of blood group A (they are all group B, AB, or O). The control
group has a larger proportion of subpopulation 2 (80%) than subpopulation 1 (20%). Thus, blood
group A frequency is 0.40 in cases and 0.10 in controls, an obvious risk association. However, when
the two subpopulations are analyzed separately, blood group A frequency is actually the same in
cases and controls from subpopulation 1, and blood group A does not even exist in subpopulation 2.
It is when the two subpopulations are mixed that a spurious difference in blood group A frequency
emerges that appears as an association between blood group A and the disease.

Genetic associations can also be modified

Effect modification is less well considered than it should be in genetic association studies.
It is a common phenomenon and very useful when considering disease biology. **Figure
9.6** shows an effect modification of an association, in this case by sex. When an association is modified by a factor (such as gender), different categories of that factor (males and
females, in the case of gender) show differences in their associations with the outcome.
The differences may be in the strength or direction of the associations, and investigators
should be aware of the possibility of masked associations that can only be unraveled by
considering potential effect modification.

In practice, it is common to see any difference observed between categories of a variable
called an effect modification, but this is not necessarily correct. For example, if odds
ratios for males and females in an association study are 1.9 and 1.1, respectively, it does
not necessarily indicate an effect modification. Whether the difference is truly an effect
modification is decided by assessing the statistical significance of the difference between
the two odds ratios, called testing for statistical interaction. Only when this analysis

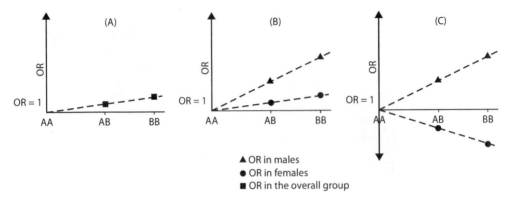

Figure 9.6 Effect modification by gender. When the analysis is repeated for males and females separately (B, C), they show differences; gender-specific associations may differ in strength (B) or even in direction (C), resulting in no statistically significant association in the overall group. See also Table 5.1.

indicates statistical significance can the difference then be called an effect modification. When this is the case, it can be concluded that the factor whose categories confer different levels of risk interacts with the genotype in question. Logistic regression allows testing for interaction on a computer. Interaction testing of a possible effect modification is a commonly omitted statistical analysis and deserves more attention than it currently receives.

The interacting factor does not have to be a nongenetic variable such as gender or ethnicity. It could be another genotype or an environmental exposure. The interpretation of an interaction between two genotypes is that genotype 1 confers risk only in the presence or absence of genotype 2. A well-replicated example in the literature is the interaction between the C282Y mutation (rs1800562) of *HFE* and the S142G variant (rs3817672) of *TFRC* in association with cancer risk. The *TFRC* variant does not modify the risk on its own, but the C282Y association becomes statistically significantly stronger in subjects homozygous for the variant allele of rs3817672. Since two variables are interacting in this example, it is known as a two-way interaction. If the interaction also involved gender (so that it only occurs in males, for example) it would be an example of a three-way interaction. As discussed in Chapter 5, in the context of gene and environment interactions, there are different types of interactions. While quantitative interactions are better recognized, unraveling them requires a very large sample size, and they are usually not examined. Qualitative interactions are easier to identify.

Confounding and effect modification may cause replication failure

If an association found in a well-designed study cannot be replicated in a second well-designed and sufficiently powered study, one possible explanation is that the first association was an indirect association with an unknown causal variant and that the same relationship with the causal variant does not exist in the sample of the second study (**Figure 9.7**A). This is termed confounding by loci and is most commonly due to a difference in the degree of LD between the two populations. Complete lack of the causal variant is also possible in extreme cases involving very distant populations. Likewise, if the first association was actually due to an unrecognized effect modification, and the modifier is rare or does not exist in the second population, the result will not be replicated (**Figure 9.7**B). For example, the *MTHFR* association with childhood leukemia is only observed in

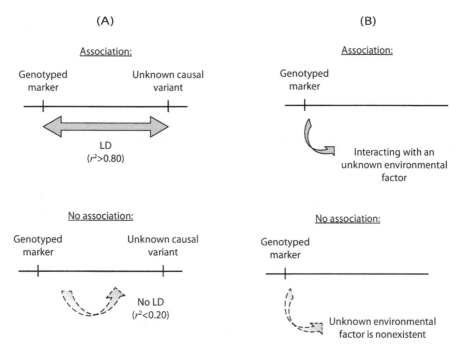

Figure 9.7 Confounding and effect modification may cause true replication failure. (A) Strong linkage disequilibrium (LD) between the genotyped marker and the unknown causal variant gives rise to the observed association in one population; the association does not occur in another study due to the lack of LD in the second population. This would have been avoided if the initial association was directly with the causal variant. (B) The marker shows an association due to an unrecognized interaction with an environmental factor, but the second population lacks that environmental factor.

populations where average folate intake is low, and the *HFE* association with childhood leukemia is exclusive to populations where iron intake is high. As replication of original findings clearly enhances the positive predictive value of the research findings being true, and findings that are not replicated will often be dismissed, it is important to recognize the reasons for genuine replication failure.

Mendelian randomization may offer protection from confounding for genetic association studies

Mendelian randomization is a natural phenomenon that occurs at conception, as each allele has an equal probability that it will be transmitted to the offspring (Mendel's second law). Since genotypes formed by alleles are the exposure in a genetic association study, this random distribution of alleles, independent from potential confounders, generates a situation close to a randomized clinical trial, where exposure is randomly (and blindly) applied to participants. Who gets which genotype is not influenced at all by external factors such as socioeconomic status, age, or gender, which are potential confounders for nongenetic studies. Family-based association studies that use parents and offspring provide the best use of Mendelian randomization and are virtually analogous to a randomized clinical trial. Due to Mendelian randomization, genetic association studies are less prone to confounding than conventional risk-factor epidemiology, although **pleiotropy** and LD still produce some confounding of genetic origin.

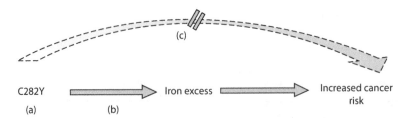

Figure 9.8 The use of the Mendelian randomization principle to test whether iron excess has a causative role in increased cancer risk. The presence or absence of the C282Y mutation in an individual (a) is not influenced by demographic features, ethnicity, socioeconomic level, or diet. The mutation is known to increase body iron levels (b) and has no other influence on body physiology (c) that may modify cancer risk. If a C282Y association with increased cancer risk is found, it can be attributed to a causative effect of iron excess without concerns over reverse causation and confounding by an environmental factor.

Mendelian randomization naturally allows the testing of whether a risk factor is likely to be on the causal pathway of disease development. This approach requires the presence of genetic markers with certain features to be used as **instrumental variables** as a proxy for the nongenetic risk factor. By examining the association between an instrumental variable and the disease, a causal inference about the role played by the nongenetic risk factor can be made, even though it has not been measured. A genetic marker can be used as an instrumental variable if it correlates with the nongenetic risk factor, which is an intermediate phenotype. Examples include the *HFE* C282Y mutation known to cause iron excess (**Figure 9.8**), *FTO* variants that increase body mass index, and a set of genetic variants that cause changes in the levels of plasma low-density lipoprotein cholesterol, plasma high-density lipoprotein cholesterol, or vitamin D. These markers can then be used (and, indeed, have been used) to investigate whether iron excess, body mass index, cholesterol, or vitamin D levels are causally associated with diseases. Likewise, lactase (*LCT*) and *ALDH2* variants that interfere with dairy and alcohol intake, respectively, may also be used as instrumental variables in Mendelian randomization studies on diseases that show correlations with dairy or alcohol intake, respectively. Since these genetic markers exclusively correlate with these nongenetic factors, genotypes can be used as proxies for lifetime increased or decreased exposures instead of measuring the exposure to nongenetic factors or their blood levels. The causal inference would be valid as long as the assumptions for being an instrumental variable are met.

For the success of the Mendelian randomization approach, the genetic markers should be the causal ones for the risk factor (or the intermediate phenotype), and should not cause any difference for nongenetic factors relevant in the biology of the disease in question. A well-understood functional genotype affecting the level of the risk factor can be used as an instrumental variable if it satisfies these assumptions.

As an example, for the C282Y mutation (see Figure 9.8) to be used in examination of the causal association between iron excess (the phenotype) and cancer risk (the outcome), the following has to be assured:

- C282Y is independent of measured and unmeasured confounders
- There is no linkage disequilibrium between C282Y and a marker that causes cancer

- C282Y is associated with iron excess

- Every pathway from C282Y to cancer passes through iron excess (that is, there is no pleiotropic effect so that C282Y is associated with cancer via a different pathway)

Iron excess is known from multiple cohort study results to increase cancer risk, but case-control studies may show an association between low serum iron levels and gastrointestinal cancer due to reverse causation resulting from iron loss via bleeding from cancer. A Mendelian randomization study is free from reverse causation, and can confirm that it is actually iron excess that is associated with cancer risk by showing an association of C282Y rather than low serum iron. Mendelian randomization is acknowledged to be an approximate rather than an absolute method, and it should also be recognized that some nongenetic characters are not always evenly distributed across genotypes.

Analysis should not be limited to single variants

Once allelic or genotypic association analyses are completed, regardless of the results, there is still more to do. In fact, restriction of the analysis to individual variants may well be the source of the current missing heritability issue. It is unlikely that individual variants act in isolation in causing a disease. More meaningful results may be obtained by using approaches that examine sets of variants acting in concert. The most intuitive of these are gene or pathway analysis, in which all of the variants of a particular gene or genes participating in the same biologic pathway are considered as a whole for joint effects. Even in the absence of an individual variant association, a pathway analysis may yield a statistically significant association signal. There are other methods that investigate **epistatic effects** of multiple variants, such as multifactor dimensionality reduction (MDR) or focused interaction testing framework (FITF), which may yield epistatic effects in the absence of individual associations. There are also special regression methods more suitable for regions of high LD, such as the HLA region.

The danger of data mining without *a priori* hypotheses is that it may turn into a data-dredging exercise that leads to spurious findings. Stringent safeguards for multiple comparisons (such as adjustment using the false discovery method; see later in this chapter) should be in place and independent replication of any positive finding should be planned. The multiple comparisons issue is the main reason why researchers are not enthusiastic about examining the millions of SNPs in a GWAS data set for pairwise interactions in order to find gene–gene interactions. It would require billions of statistical tests—which is no longer an issue with the computational power now available—but increasing the number of tests increases the likelihood of false-positive findings. The current trend is to examine only the interactions that have been hypothesized *a priori*, or to run an exploratory interaction analysis as a secondary approach.

Data from genetic association studies can be used for further research

Data, especially the healthy controls data, can be used for further population genetics analyses, indications of natural selection, assessment of age- and sex-based genotype frequency differences, and for phylogenetics analysis. Even if the researchers involved in the genetic association study are unable to do any more analyses on their data, others can. To enable other researchers to benefit from existing data, the sharing of data, with adherence to regulatory norms, should be considered. Any well-designed study is valuable, even if the results are negative, as it can contribute to a future meta-analysis. Some journals now have sections for negative results, and there are journals that welcome reports without a strong finding as long as they are well designed and technically sound. To minimize publication bias and to maximize the chances of being included in a meta-analysis, every effort should be made to report the study and its results.

9.2 GWAS Analysis

Minor errors in GWAS design can cause large errors in the data

With the large sample size usually consisting of multiple batches, and millions of variants, the approach to GWAS data is very different from that used for smaller-scale candidate gene studies. GWASs may use highly advanced technology for genotyping and the most up-to-date statistical methods for data analysis, but nothing can replace a good study design. The impact of even minor amounts of genotyping errors (Chapter 7), population stratification (Chapter 3), or cryptic relatedness (Chapter 3) among participants is disproportionately large in a GWAS. This is why even well-designed and executed GWASs are routinely subjected to the most stringent quality control measures (**Table 9.1**).

Most of the quality control steps included in Table 9.1 aim to avoid any possible bias introduced by (a) differences in the quality of DNA samples from subgroups of case-control

Table 9.1 Quality control filters before GWAS data analysis

Level of quality control	Sample (cases and controls)	SNPs genotyped	Data (genotypes)
Filters or tests (aim and/or filter value)	**Missingness** (>97% of SNPs should have results for each subject)	**Missingness** (each SNP should have results in ≥98% of cases and controls)	**Quantile-quantile (QQ)** plots (to visualize systematic errors)
	Gender checks using X-chromosome data	**Minor allele frequency** (≥0.01; very rare alleles require specialized analysis methods)	
	Duplicates (no exact duplicates can be included. Exclude the duplicate with higher missingness values)	**HWE testing** ($P < 10^{-4}$)	
	Cryptic relatedness (no two or more samples should appear to be closely related)	**X-chromosome-specific tests** (heterozygosity for X-chromosome SNPs is possible in females but not in males)	
	Ethnicity check (by principal component analysis)	**Visual inspection of signal intensity plots** (checks for genotyping quality and being in a CNV region)	
	Heterozygosity (for each subject, it should be within the mean ±3 standard deviations over all samples)	**Difference between control groups, if more than one** ($P > 10^{-4}$ for the additive model)	
		Sex-specific frequency difference within the control group ($P > 10^{-4}$ for the additive model)	

Samples 5763
• 18 duplicated samples excluded
• 219 samples excluded with call rate <97.5%
• 11 samples excluded with excess of heterozygosity
• 40 samples excluded of non-Caucasian origin
• 170 samples excluded with familiar relationships
• 61 samples excluded with gender mismatch
• Final 5244 samples

SNPs 657,366
• 95,876 probes filtered for CNV intensity
• 4298 SNPs excluded with call rate <95%
• 557,192 SNPs left for analysis

Figure 9.9 A typical quality control process for the samples and SNPs in a GWAS. Samples were excluded for the presence in duplicates, non-Caucasian origin (as identified by genetic analysis), a call rate lower than the predetermined threshold, an autosomal heterozygosity rate in excess of the predetermined threshold based on the overall sample average, cryptic relatedness, and gender mismatch. A large number of SNPs failed the quality control test for not meeting the call-rate threshold or for being identified as within CNV regions. (Adapted from Trompet S, de Craen AJ, Postmus I et al. [2011] *BMC Med Genet* 12,131. With permission under the terms of the Creative Commons Attribution License.)

samples (the worst-case scenario is that case and control DNA samples are from different sources like blood and saliva, are extracted using different methods, and are stored differently) and (b) differences in genotyping quality due to the genotyping of subsets of the sample in different locations, by different methods, or at different times. Genotyping quality may be different from plate to plate, instrument to instrument, day to day, or among batches of DNA samples. This is why a major principle of experimental design—blocking what you can and randomizing what you cannot—also applies to genotyping. The quality steps assure that the sample and SNP properties do not introduce any bias into the results. A consortium meeting in Travemünde in 2007 agreed on a set of quality control filters known as the **Travemünde criteria**, which are incorporated in Table 9.1. Applying these filters is now a routine practice in GWAS analysis and these steps are implemented in major data analysis packages (for example, PLINK). **Figure 9.9** summarizes the quality control procedures applied to a real GWAS.

Systematic errors in GWAS data are explored using a QQ plot

Systematic errors that may result in spurious associations are explored using a graphic method called a quantile-quantile (QQ) plot. To construct this plot, all observed P values are ranked from the smallest (the most significant) to the largest and plotted against the expected null distribution (**Figure 9.10A**). Each P value is converted to its negative log value. Thus, $P = 10^{-8}$ becomes +8. Each observed P value is paired with an expected value from a theoretical Chi-squared (χ^2) distribution and calculated as:

$$-\log(i/(L+1))$$

where i is the rank of the P value and L is the total number of SNPs. Thus, if the top hit has a P value of 10^{-8} in a GWAS that included 1,000,000 SNPs, the pair of numbers will be (8, 6). Likewise, in a study of 10,000 SNPs, if the fiftieth most significant P value is 10^{-5}, then the pair of (x, y) values will be (5, 2.3) since $-\log(50/10,001)$ is 2.3. These pairs of numbers are then plotted in a graph. If there is no association, the plot of the pairs of numbers will form a straight line along the 45° reference line. Deviations from the $y = x$ line correspond to loci

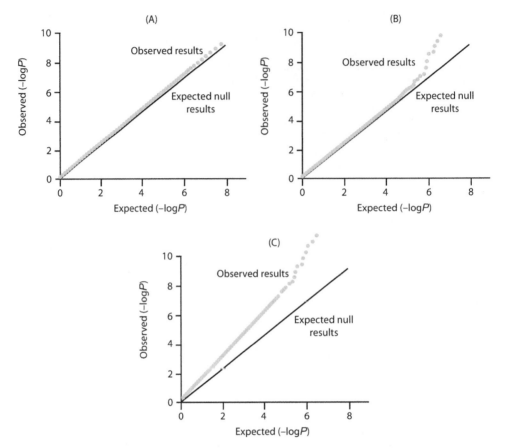

Figure 9.10 Examples of QQ plots visualizing observed GWAS results against theoretical expectations. The expected null distribution (the identity line) is a straight line. (A) No associations: the expected and observed results show similarity. Only an expected number of results are statistically significant given the number of comparisons made. (B) A typical good GWAS result. For the majority of SNPs, there is no more statistical significance than would be expected when this many SNPs are examined, but for a number of SNPs, statistical significance values are higher than expected, indicating true associations. (C) A problematic result showing a generalized shift of observed results from expected null results for any level of statistical significance. Such a generalized shift indicates a systematic error such as population stratification, and these associations are more likely to be spurious.

that deviate from the null hypothesis (**Figure 9.10B**). If there is a systematic shift from the $y = x$ line, beginning right from the origin, this is usually an indication of population stratification causing systematic spurious associations (**Figure 9.10C**). Systematic genotyping errors can also generate a similar shift.

The shape of the curve and its relationship with the identity line reveal a lot of information. The nature of the deviation from the identity line indicates whether the observed associations are likely to be true or due to population stratification, cryptic relatedness, or some other systematic error. The protocol listed in the Further Reading section provides detailed guidance on how to produce a QQ plot.

Statistical analysis of GWAS data uses specialized software

GWAS data analysis is routinely performed for each SNP individually. It is not essentially different from what is done in a candidate gene study, but the large number of SNPs to be analyzed creates a major task and requires some computing power. Special programs have been devised to handle the analysis of millions of SNPs in a GWAS. The most popular of these programs is PLINK, which is a free, open-source, whole-genome association analysis toolset designed to perform a range of basic, large-scale analyses in a computationally efficient manner. PLINK is able to perform association tests for each genetic model by the Chi-squared or Fisher's exact tests, and can also perform the permutation test. GWAS results are usually presented as **Manhattan plots (Figure 9.11)** for which data have to be exported into a program such as Haploview. All of the quality control tests, statistical data analysis, and production of the QQ plots and Manhattan plot can be completed within hours by experienced users. Additional capabilities of PLINK include haplotype-based association analysis, imputation, interaction analysis, and meta-analysis. For population stratification assessment, PLINK uses a distance measure based on genome-wide pairwise identity-by-state. Although it is not as powerful as principal component analysis (explained later in the chapter), it has been used successfully in a number of studies. PLINK also provides an option to include **genomic control (GC)** adjustment in the analysis. To generate a QQ plot, PLINK provides a file with results that are exported to the

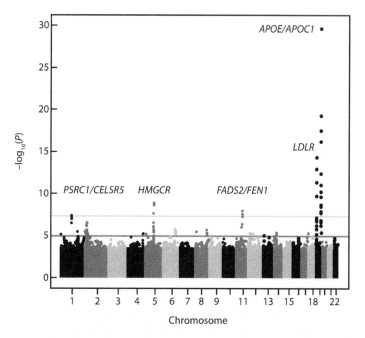

Figure 9.11 An example of a Manhattan plot summarizing GWAS results. This study investigated genome-wide associations with blood low-density lipoprotein levels. The top hit is a SNP in the *APOE* gene with a *P* value of 10^{-30}. Other associations exceeding the statistical significance threshold are also shown. The higher horizontal line is the consensus threshold for genome-wide statistical significance (5×10^{-8}), and the lower line corresponds to $P = 10^{-5}$, which is the threshold favored by proponents of Bayesian analysis. (Adapted from Trompet S, de Craen AJ, Postmus I et al. [2011] *BMC Med Genet* 12, 131. With permission under the terms of the Creative Commons Attribution License.)

R package. The protocol listed in the Further Reading section provides step-by-step guidance to doing the analysis (including quality control tests) and on how to generate Manhattan and QQ plots on Haploview and R, respectively. The PLINK Website also provides extensive guidance on its use.

The statistical significance threshold is different for a GWAS

The main statistical issue with GWAS analysis is the genome-wide statistical significance threshold, which is set at 5×10^{-8} and is the Bonferroni-corrected genome-wide type I error rate of 0.05. This is considered to be too conservative by some researchers and is given as the reason for the high false-negativity rate produced by a GWAS. It has been proposed that, according to Bayesian analysis principles, the genome-wide statistical threshold of $P = 10^{-5}$ is sufficient to prevent false positives. Adoption of a strict threshold would prevent false-positive results but at the expense of potentially increasing the false-negative rate. **Bonferroni correction** is not the only method used for prevention of type I errors. Other popular approaches are briefly discussed below.

Population stratification must be assessed in each GWAS

Population stratification is the allele frequency differences between cases and controls due to ancestry differences, and is recognized as a major cause of spurious associations (Chapter 3). Population stratification is an important issue for any genetic association study, but it is exceedingly so for GWASs due to the large sample size and the pooling of multiple population samples almost as a routine practice. Ironically, a large sample size exaggerates the effects of population substructure on association study outcomes. Admixed populations such as Hispanic and African American are most vulnerable to population substructure. These populations contain a mixture of genetic material from different ancestral populations and the proportions vary in each individual. Such populations require additional attention. For example, one study used 116,204 SNPs to estimate the genetic ancestry of 96 Puerto Rican participants who identified themselves, their parents, and grandparents as Puerto Rican. When their ancestry proportions were determined using these SNPs and genotype data from HapMap ancestral populations, substantial heterogeneity in ancestry among individuals was found. The proportions of African ancestry showed the widest range (from less than 10 to over 50%). European ancestry proportions in these 96 individuals ranged from under 20 to over 80%. The third category, Native American ancestry, showed proportions in their genomes ranging from 5 to 20%. It is, therefore, possible to estimate proportions for each ancestry's contribution to the individual genomes and to use these values to cluster individuals in subpopulations. Such clusters determined by molecular markers are more robust than those identified by geographic and ethnic data.

There are a number of approaches to control confounding by ethnicity and all require genome-wide SNP data and computing power. Note that the population structure being ruled out is not an obvious one, as shown in Figure 9.5, but a more cryptic population structure. Even a sample of all white Europeans or all Africans may have a degree of cryptic population structure, which needs to be assessed. The most popular methods are briefly discussed here, but their details are beyond the scope of this book.

Genomic control is the simplest of all methods to control for population stratification. This method was originally devised for candidate gene studies and used 10 to 50 neutral markers. It seeks to lower the false-positivity rate by increasing the statistical threshold to a degree determined by a factor estimated from the data. Simply, if loci that are neutral for disease susceptibility but show frequency differences among the populations included in the study show no association with the phenotype in a well-powered study, then it may be concluded that population stratification is not a great problem. This is the minimum a candidate gene study should aim for in order to rule out a major contribution of population stratification to the obtained results. In a GWAS, thousands of neutral loci, called genomic controls, are available and are used to correct for stratification by adjusting association statistics. The correction factor is a uniform overall inflation factor (γ) that is calculated from the association statistics of these neutral markers. The correction is performed by dividing the P value for each SNP association by this value (which is usually in the range 1.04 to 1.76; the GWAS analysis illustrated in Figure 9.10 yielded a very small γ value of 1.077). Dividing the P values by γ provides genomic-control-adjusted P values. The disadvantage of this method is the uniform application of the inflation factor to all SNPs regardless of the magnitude of their own frequency differences among ethnicities. However, if the overall inflation factor is not large, it may be seen as a reassurance against major population stratification. Further, confounding may also result in false negatives and the GC method does not do anything about this.

Structured analysis (SA) uses clustering algorithms to define subpopulations within the population sample used in the study, and estimates the number of clusters. It does this by inferring subpopulation memberships for each individual. The next step is to conduct a test of association within each subpopulation. The overall results are then statistically adjusted using the structure in cases and controls. The structuring is done on theoretical grounds and, for GWAS data, it is highly computationally demanding.

Principal component analysis (PCA) is a popular statistical method used in many areas as a nonparametric method to extract relevant information from large, confusing data sets. This is currently the most popular method for the assessment of population substructure and uses SNPs genotyped in a GWAS. PCA is performed on EIGENSTRAT, which is an open-access program. The advantage of PCA is that the resulting correction is specific to the SNP's variation in frequency across ancestral populations. This method has been found to be the most sensitive for detection of within-continent genetic variation and is computationally less demanding than other methods. PCA is now the gold standard for population substructure assessment in a GWAS and the results are used to make necessary adjustments of the study results. As the authors of the original study using EIGENSTRAT pointed out, although EIGENSTRAT is a robust and powerful method for correcting for stratification, it is not a panacea, and the principles of careful experimental design and the matching of the ancestry and laboratory treatment of cases and controls to the fullest extent possible should be adhered to.

Multiple comparisons are critical in the analysis of millions of variants

A GWAS analysis inevitably includes millions of comparisons of genotype frequencies, usually between cases and controls. The problems associated with multiple comparisons are well recognized. A large proportion of P values resulting from multiple comparisons may indicate false-positive associations and some adjustment is needed. The simplest correction for multiple comparisons is the classic Bonferroni correction, which involves multiplication of the obtained P value by the number of comparisons performed. For example, if the P value is 0.001 and 100 statistical tests have been performed, the corrected P value becomes $0.001 \times 100 = 0.1$ and loses statistical significance. The Bonferroni adjustment works best if all SNPs are independent (no LD among them), but this assumption is rarely met in genetic studies. Thus, it is usually considered a conservative method, meaning it exaggerates the correction and may cause false negatives. The currently used genome-wide statistical significance threshold is based on the Bonferroni correction for one million SNPs, and is likely to err on the conservative side.

Permutation testing provides a very different approach to the corrections for multiple comparisons. First, the standard statistical test is performed and a P value is obtained (the original P value). The permutation test then shuffles the case-control status of each sample randomly and then runs the association analysis. This process is repeated thousands of times (permutations) as determined by the user. The P values estimated in each permutation are retained. The distribution of permutation P values is compared with the original P value. This comparison leads to an estimate of how often the original P value would occur by chance or randomly if the study was repeated many times. This estimate is obtained by checking the percentage of permutation P values that are smaller than the original P value. This percentage is the P value for the permutation test. Since each permutation on each SNP assesses the role of chance, the results are free from the bias of multiple comparisons, and this can be applied to as many SNPs in the study as one wishes without needing further correction. While the method looks attractive, the computing demands are high for a GWAS and it is not frequently used. Current computers are capable of handling this task, although it takes a longer time than analysis by logistic regression using PLINK.

The approach that is becoming more popular is the **false discovery rate (FDR)** method originally described by Benjamini and Hochberg. This is a less stringent form of adjustment than the Bonferroni adjustment for the number of comparisons made. It quantifies the false discovery problem, and the resulting FDR is the estimated proportion of false positives among those tests that are deemed significant. The FDR procedure calculates a q value, which indicates the proportion of statistically significant results obtained by the original test procedure that are false positives. Following a ranking procedure, the P values ranking below the threshold determined by the user-determined q value are considered to indicate true positives that are free of the potential false positives resulting from multiple comparisons. Corrected P values can be calculated by multiplying each P value by the total number of comparisons made and dividing this value by the rank of the P value (the smallest is ranked 1). This procedure is receiving more acceptance than the more stringent Bonferroni correction, and probably results in fewer false negatives.

There are good reasons for being cautious when interpreting association results

All genetic association studies are observational studies, and GWASs are hypothesis-generating studies. Such studies are subject to unintended biases that distort the results. While false-negative results are possible, the field has suffered more from false positives and some extreme measures have had to be taken to prevent more of them. The pitfalls and limitations of genetic association studies must be recognized by every researcher, and the value of replication to validate results must be appreciated. A healthy dose of skepticism in the interpretation of association study results is very useful and alternative explanations must be considered. The most common alternative explanations for genetic association study results are chance, confounding by locus, confounding by ethnicity, and bias. It has to be remembered that no statistical adjustment can fix confounding or bias. A true replication study goes a long way to confirm the findings before considering biomarker development or any intervention, which may be a drug development project or a public health policy change.

Reporting genetic association study results

With the increasing popularity of all kinds of genetic association studies, it has become obvious that certain standards have to be established for publication of results, much like the existing standards for reporting biomarker, microarray, or quantitative PCR results. The existing STROBE (Strengthening the Reporting of Observational Studies in Epidemiology) statement has been tailored for genetic association studies and is called the STrengthening the REporting of Genetic Association Studies (STREGA) statement. There is also a Strengthening the Reporting of Molecular Epidemiology for Infectious Diseases (STROME-ID) statement. STROBE and STREGA statements are accessible in open-access journals and also via relevant Websites (see the URL list at the end of the chapter).

Key Points

- Well-designed and well-powered genetic association studies are particularly powerful tools to probe disease biology as they are not influenced by reverse causation.

- Major pitfalls that need to be addressed in genetic association study analysis include genotyping error, population substructure, and multiple comparisons.

- Well-powered, large studies with incorporated replication arms, that exclude population substructure as an alternative explanation and take into account the multiple comparisons issue, are more likely to show true associations rather than generate spurious findings.

- Mendelian randomization can help with strengthening causal inference in certain circumstances.

- As for any observational study, genetic associations may suffer from chance findings, confounding by locus or ethnicity, and bias.

- Each genetic association study result is obtained from an observational study and, regardless of the advanced technology or statistical method used, they require independent replication for validation.

- Quality control steps routinely applied to GWAS data analysis are no substitute for good practice in the design and execution of a study.

- Multiple comparisons must be taken into account in the analysis and interpretation of genetic association study results. The Bonferroni adjustment is not the only method to correct for multiple comparisons.

- Current GWAS results are individual SNP associations in the overall sample. Sex-, age-, and population-specific associations and those involved in epistatic associations without main effects are missing.

- The resolving of observational correlations into causal relationships is not easy and delays the understanding of disease biology, disease prevention, drug development, and clinical translation.

URL List

DeFinetti. Online HWE and Association Testing. http://ihg.gsf.de/cgi-bin/hw/hwa1.pl

EIGENSTRAT (EIGENSOFT). http://genetics.med.harvard.edu/reich/Reich_Lab/Software.html

Haploview. Broad Institute. http://www.broadinstitute.org/scientific-community/science/programs/medical-and-population-genetics/haploview/haploview

PHASE. Department of Human Genetics. Department of Statistics. University of Chicago. http://stephenslab.uchicago.edu/software.html#phase

PLINK. http://pngu.mgh.harvard.edu/~purcell/plink and http://pngu.mgh.harvard.edu/~purcell/plink/plink2.shtml

R Project for Statistical Computing. http://www.r-project.org

STREGA Statement. University of Ottawa. http://www.medicine.uottawa.ca/public-health-genomics/web/eng/strega.html

STROBE Statement. University of Bern. http://www.strobe-statement.org

UNPHASED. https://sites.google.com/site/fdudbridge/software/unphased-3-1

Further Reading

Anderson CA, Pettersson FH, Clarke GM et al. (2010) Data quality control in genetic case-control association studies. *Nat Protoc* 5, 1564–1573 (doi: 10.1038/nprot.2010.116). (*Provides a step-by-step protocol for data quality tests for candidate gene studies and GWASs that can be performed by PLINK. It also details the use of principal component analysis in the assessment of population stratification.*)

Attia J, Ioannidis JP, Thakkinstian A et al. (2009) How to use an article about genetic association: B: Are the results of the study valid? *JAMA* 301, 191–197 (doi: 10.1001/jama.2008.946). (*The second of a three-piece review series, this one covers the validity of genetic association study results. Hardy–Weinberg equilibrium, population substructure, and genotyping errors are among the topics covered.*)

Ball RD (2011) Experimental designs for robust detection of effects in genome-wide case-control studies. *Genetics* 189, 1497–1514 (doi: 10.1534/genetics.111.131698). (*This paper provides a good insight into a lot of statistical issues related to GWAS analysis, including the Bayesian approach as opposed to the classic (frequentist) statistical analysis. A good source for those interested in Bayesian analysis, which is not covered in this book.*)

Bush WS & Moore JH (2012) Chapter 11: Genome-wide association studies. *PLoS Comput Biol* 8, e1002822 (doi: 10.1371/journal.pcbi.1002822). (*A very accessible paper reviewing key concepts including the background to genetic association studies, human genetic variation, GWAS platforms, study designs, and the statistical methods used for data analysis.*)

Clarke GM, Anderson CA, Pettersson FH et al. (2011) Basic statistical analysis in genetic case-control studies. *Nat Protoc* 6, 121–133 (doi: 10.1038/nprot.2010.182) (*A step-by-step protocol for basic statistical analysis in a population-based genetic association case-control study using PLINK, supplemented by Haploview and R package. It also provides formulas for the calculation of odds ratios, Chi-squared values, and test statistics for the Cochrane–Armitage test. The derivation of the Q-Q plot, Manhattan plot, and PL plot is explained too.*)

Cordell HJ & Clayton DG (2005) Genetic association studies. *Lancet* 366, 1121–1131 (doi: 10.1016/S0140-6736(05)67424-7). (*This review paper is written by eminent statisticians for general readers. It outlines the main principles of genetic association studies, including most of those covered in Chapters 8 and 9 of this book, as well as other topics not covered in this book.*)

Davey Smith G & Hemani G (2014) Mendelian randomization: genetic anchors for causal inference in epidemiological studies. *Hum Mol Genet* 23(R1), R89–R98 (doi: 10.1093/hmg/ddu328).

Huckins LM, Boraska V, Franklin CS et al. (2014) Using ancestry-informative markers to identify fine structure across 15 populations of European origin. *Eur J Hum Genet* 22, 1190–1200 (doi: 10.1038/ejhg.2014.1). (*A nicely presented example of the use of PCA, which makes the PCA concept very clear.*)

König IR (2011) Validation in genetic association studies. *Brief Bioinform* 12, 253–258 (doi: 10.1093/bib/bbq074). (*After a general review of genetic association studies, this paper emphasizes the importance of replication and validation, including causality assessment, in the identification of genetic variants responsible for complex diseases.*)

Lambert CG & Black LJ (2012) Learning from our GWAS mistakes: from experimental design to scientific method. *Biostatistics* 13, 195–203 (doi: 10.1093/biostatistics/kxr055). (*Presents the senior biostatisticians' approach to a GWAS and how past mistakes can be used to learn for the future. As a paper providing a critique of the GWAS as a scientific method, it is invaluable reading for beginners.*)

Little J, Higgins JP, Ioannidis JP et al. (2009) STrengthening the REporting of Genetic Association Studies (STREGA)—An Extension of the STROBE Statement. *PLoS Med* 6, e1000022 (doi: 10.1371/journal.pmed.1000022). (*A consensus statement by the pioneers of the field on what and how to report for genetic association study results. For each recommendation, background information is provided, which makes this paper an educational resource. Adherence to these recommendations is being forced by most journals.*)

Lunetta KL (2008) Statistical Primer for Cardiovascular Research: Genetic association studies. *Circulation* 118, 96–101 (doi: 10.1161/CIRCULATIONAHA.107.700401). (*Another easy-to-follow review of genetic association studies; it is particularly useful for study designs, genetic models, and multiple comparisons and provides an example of an association study analysis.*)

Montana G (2006) Statistical methods in genetics. *Brief Bioinform* 7, 297–308 (doi: 10.1093/bib/bbl028). (*This review covers statistical background to genetic association data analysis as well as some of the advanced topics not included in this book. These include multimarker analysis, admixture mapping, and data-mining methods for interaction detection.*)

Neale B, Ferreira M, Medland S & Posthuma D (eds) (2007) Statistical Genetics: Gene Mapping Through Linkage and Association. Taylor & Francis. (*This handbook for statistical genetics is a comprehensive review of concepts and methods covering methodologies that are beyond the scope of this book. Readers requiring information on more advanced topics will find it very useful.*)

Pearson TA & Manolio TA (2008) How to interpret a genome-wide association study. *JAMA* 299, 1335–1344 (doi: 10.1001/jama.299.11.1335). (*An overview for general readers on study designs, selection of cases and controls, genotyping and quality control, analysis and presentation of results, replication and functional studies, and the limitations of a GWAS. It also provides ten basic questions to ask about a genome-wide association study report.*)

Trompet S, de Craen AJ, Postmus I et al. (PROSPER Study Group) (2011) Replication of LDL GWAs hits in PROSPER/PHASE as validation for future (pharmaco)genetic analyses. *BMC Med Genet* 12, 131 (doi: 10.1186/1471-2350-12-131). (*A typical GWAS report. The examples of a Manhattan plot, QQ plot, and quality control procedure provided in this chapter have been reproduced from this paper.*)

Turner S, Armstrong LL, Bradford Y et al. (2011) Quality control procedures for genome-wide association studies. *Curr Protoc Hum Genet* Chapter 1: Unit 1.19 (doi: 10.1002/0471142905.hg0119s68). (*A GWAS quality control review paper with illustrations from real data sets.*)

Weale ME (2010) Quality control for genome-wide association studies. *Methods Mol Biol* 628, 341–372 (doi: 10.1007/978-1-60327-367-1_19). (*A comprehensive review of quality control steps in a genetic association study with accompanying protocols.*)

Wellcome Trust Case Control Consortium (2007) Genome-wide association study of 14,000 cases of seven common diseases and 3,000 shared controls. *Nature* 447, 661–678 (doi: 10.1038/nature05911). (*This is the first true GWAS that implemented most of the modern methods that were then used in following studies. This 18-page Nature paper provides plenty of details about how a GWAS is run and analyzed. The supplementary information file (http://www.nature.com/nature/journal/v447/n7145/extref/nature05911-s1.pdf) provides further details of each method used in the analysis (quality control and association*

*analysis) and includes background information
to each, with many illustrations, in the 56-page
document.)*

Ziegler A (2009) Genome-Wide Association
Studies: Quality control and population-
based measures. *Genet Epidemiol*
33(Suppl 1), S45–S50 (doi: 10.1002/
gepi.20472). (*One of the best reviews
available for quality measures in GWAS
analysis detailing the Travemünde criteria.*)

Bioinformatics for the Interpretation of Genetic Association Study Results

10

Genetic association studies can provide **statistically significant** results but further studies are required to establish biological significance. Functional replication studies on genetically replicated results aim to confirm or refute whether the statistical correlation between the variant and disease risk points toward a potentially causal relationship. Recent advances in computational **bioinformatics**, as well as the availability of massive amounts of genomics and proteomics data, enable researchers to carry out computational analyses in a dry laboratory to pinpoint the functional variants. The ultimate aim is to treat the statistical result as an association signal to find out the actual causal variant. This chapter will present the bioinformatics tools and approaches available to take association studies a step further. Bioinformatics is a very broad and ever-expanding discipline. Here, only what is emerging as **translational bioinformatics** is covered, with a focus on the functional assessment of genetic variants.

10.1 Bioinformatics for Testing SNP Associations

It is common practice that association studies are reported with a P value, which is usually accompanied by an effect-size estimate. Although the current standard is to incorporate a replication study into the main project, functional follow-up is less common. One of the main drawbacks of the current GWAS approach is that most results are indirect associations. Unlike a direct association of a causal SNP with disease susceptibility, an indirect association is with a SNP that is in linkage disequilibrium (LD) with the causal SNP. A comprehensive analysis of autoimmune disease-associated SNPs suggested that SNPs reported in the GWAS catalog have on average a 5% chance of representing a causal SNP that is functional and therefore affects the function of a gene or a number of genes. For efficiency, usually only SNPs that represent a group of correlated SNPs (SNPs in LD with each other) are included in GWAS chips and a detected association may therefore have originated from any SNP in the group. This group is known as a set of **statistically similar SNPs** (**ssSNPs**) and the SNP that represents them is said to be a **tagging SNP**. Members of the ssSNP set are potentially causal SNPs, but the causal SNP may also be an unknown one. Bioinformatics can examine both possibilities and help to identify the mechanism that underlies disease development (**Figure 10.1**).

Generating a ssSNP set using HapMap and 1KG data

The HapMap Project was originally conceived to generate a haplotype map of the human genome in four major populations (later expanded to 11 populations) and only used known SNPs with a minor allele frequency >0.05 to genotype population samples. The HapMap Project generated genotype data, allele and genotype frequencies, LD data, and haplotype phase information. These data are publicly available and the LD data can be used to obtain ssSNP sets for SNPs analyzed in HapMap. The software **Haploview** is specifically designed to

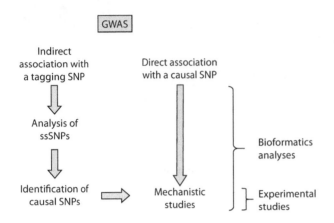

Figure 10.1 The role of bioinformatics in the post-GWAS phase of an association study.

analyze HapMap data and is also publicly available. Data from any region of the genome and any population can be downloaded to Haploview to generate an LD map and lists of ssSNP sets for tagging SNPs. The SNP of interest can then be assessed as to whether it is a tagging SNP and, if so, the list of tagged SNPs is obtained. It may be that the SNP of interest is tagged by another one and belongs to the ssSNP set for the tagging SNP. Any SNP in the ssSNP set could be the causal SNP responsible for the association signal. For any given SNP, different sets of ssSNPs may be generated using different thresholds of the LD parameter r^2. While the r^2 threshold for strong LD is 0.80, recent studies have shown that causal SNPs may be in weaker LD ($r^2 \sim 0.50$) with the tagging SNP. Thus, generating a ssSNP set using an r^2 threshold of 0.40 may be a reasonable approach. Once the ssSNP set is available, it is subjected to a process of assigning functions to the SNPs (**functional annotation**) to identify potentially causal SNPs. Although Haploview cannot analyze a genomic region larger than 500 kb, it is possible to examine LD between SNPs farther away from the lead SNP, but this requires manual manipulation of HapMap data by experienced users.

While HapMap data can be used to generate a ssSNP list, this list will only include SNPs that are genotyped in HapMap, which are usually common SNPs. The 1000 Genomes (1KG) Project has gone further than HapMap and sequenced the whole genomes of more than 1000 people from 26 populations, including the original HapMap samples. These data are freely accessible and an LD map can be generated for any part of the genome on the 1KG browser. The analysis of both HapMap and 1KG data to generate a ssSNP set for a given SNP is relatively straightforward but may be a daunting task for inexperienced researchers lacking training in the use of large genomics data sets. Fortunately, there are several online tools that interrogate the HapMap and 1KG data to produce ssSNP lists for most commonly studied populations (**Table 10.1**).

The prototype online tool for ssSNP set generation is HaploReg, which incorporates functional annotations for noncoding SNPs in the output. For single SNP analysis, the output can be changed according to the population choice and desired r^2 threshold. An example of the on-screen output is shown in **Figure 10.2**. The output includes basic descriptive information on the ssSNPs (in LD with the query SNP) and a detailed functional annotation. The annotations show whether the SNPs in the ssSNP set coincide with **histone marks** associated with promoters or **enhancers**, **DNase hypersensitivity sites** (indicating transcriptionally active sites), or transcription factor binding sites, based on empirical data. The annotation also includes data on whether the SNP is known as an eQTL for any cell or

Table 10.1 Primary online tools that can be used to generate statistically similar SNP lists

Resource	Source data	Input	Features
HaploReg	1KG	Web entry; text file for multiple SNPs; GWAS results for phenotypes from GWAS catalog	Focuses on noncoding variants
			Provides information on representation of SNPs on GWAS microarrays
			Choice of r^2 thresholds (from 0.2 to 1.0) and population (but one population at a time)
			Provides functional annotation based on ENCODE and Roadmap data
			Accepts existing GWAS results as entry
			Provides genomic coordinates for the latest two assemblies hg19 and hg38
			Output can be saved as a text file
VaDE (VarySysDB Disease Edition)	1KG	Web entry (single SNP)	Screens up to 3000 SNPs on each side of the lead SNP
			Uses essentially the same data sources as HaploReg for annotation
			Generates a visual output and provides details for each SNP in a separate pane
			Uses an r^2 threshold of 0.80 (cannot be changed)
			Shows r^2 results from European, Asian, and African populations simultaneously
SNAP (SNP Annotation and Proxy search)	1KG (phase 1) or HapMap	Web entry; text file; genomic locus	Provides information on representation of SNPs on GWAS microarrays
			Choice of r^2 thresholds and population
GLIDERS (Genome-wide LInkage DisEquilibrium Repository and Search engine)	HapMap (phases 2 and 3)	Web entry	Only SNPs with minor allele frequency >0.05 are included in the analysis
			Provides information on representation of SNPs on GWAS microarrays
			Choice of r^2 thresholds (from 0.3 to 1.0)

Web addresses are provided in the URL list at the end of the chapter.

tissue type and whether the SNP alters any transcription factor binding site (although not necessarily with experimental confirmation). HaploReg also offers information on whether a SNP alters the driver nucleotides of an enhancer region, which can be up to 600 bp long. While HaploReg extracts and presents biological data from the ENCODE project, it does not rank the ssSNPs based on their involvement with biological processes. To be able to rank the ssSNPs from the most functional (and therefore likely to be causal) to the least functional, the ssSNP list must be transferred to another resource called **RegulomeDB**, which can rank noncoding SNPs for their functionality.

As the HaploReg output shows, the results on epigenetic marks and regulatory DNA elements are given per cell, tissue, or organ. This is particularly important since most processes in the genome show specificity to the target tissue. With the availability of ENCODE and Epigenomics Roadmap data from several hundred cell or tissue types, a dry laboratory *in silico* approach that does not demand as much resource in terms of funds, labor, and time is gaining advantage over wet laboratory experiments, which are usually restricted to one or a few cell types.

Query SNP: rs3117582 and variants with r² >=1

pos (hg 19)	pos (hg 38)	LD LD (r²)(D')	variant	Ref	Alt	AFR freq	AMR freq	ASN freq	EUR freq	Promoter histone marks	Enhancer histone marks	DNAse	Proteins bound	Motifs changed	GENCODE genes	dbSNP func annot
chr6: 31596138	chr6:31628361	1 1	rs3132450	A	G	0.04	0.02	0.00	0.07		SKIN		POL2, POL2B	Cdx, Pbx3	PRRC2A	intronic
chr6:31620520	chr6:31652743	1 1	rs3117582	T	G	0.04	0.02	0.00	0.07	24 organs	HRT		42 bound proteins	E2F, Zfp 161, Znf143	APOM	intronic
chr6:31624864	chr6:31657087	1 1	rs3117581	A	G	0.04	0.03	0.00	0.07	LIV	8 organs	53 organs		ATF3, THAP1	APOM	intronic
chr6:31626013	chr6:31658236	1 1	rs3132449	C	T	0.04	0.03	0.00	0.07	BLD	22 organs			ATFJ, Hand, Smad3	25bp 3□ of APOM	
chr6:31636742	chr6:31668965	1 1	rs9267531	A	G	0.04	0.03	0.00	0.07		6 organs	BLD		CTCF, Rad21	CSNK2B	intronic
chr6:31671557	chr6:31703780	1 1	rs9267544	C	A	0.04	0.02	0.00	0.07	22 organs	6 organs		POL2	HEN1, Nr2f,2, Rad21, p300	LY6G6E	
chr6:31677391	chr6:31709614	1 1	rs9267549	G	A	0.05	0.03	0.01	0.07			9 organs		ERalpha-a, VDR	LY6G6F	intronic

Figure 10.2 The output of a typical HaploReg analysis. This analysis is for the SNPs in absolute LD with the lung cancer-associated SNP rs3117582. Six SNPs are listed and functional annotation (epigenetic marks and ENCODE elements) is included. These results are for the European population and can be repeated for other populations.

An unknown SNP may be assessed in a dry laboratory

A statistical re-examining of associations with imputed SNPs after a GWAS can find an association without genotyping. An imputed SNP is usually a rare SNP that was not geno-typed in the original study; if it is in LD with the SNP that has provided the association signal, it is potentially a causal SNP like any other SNP in LD with the tagging SNP. If an association with an imputed rare variant in LD with the original SNP associated with the trait yields a greater statistical significance, the original association was apparently an indirect one with a tagging SNP. This statistical result still requires functional validation to gain translational value. The imputation approach has been used to re-examine existing associations in some common diseases, and some associations have been refined at the statistical level. For example, α_1-antitrypsin deficiency leading to emphysema was origi-nally associated with the deleterious missense mutation in *SERPINA1*, SNP rs28929474. After missing genotypes were inferred by imputation and the analysis was re-run, the rare (minor allele frequency = 0.007) intergenic SNP rs112635299, 4.9 kb 3' downstream from *SERPINA1*, yielded a stronger association than the original observation. These two SNPs are 6 kb apart on Chromosome 14. The HaploReg analysis using rs28929474 as the tagging SNP indeed shows rs112635299 in LD with it ($r^2 = 0.78$). Imputation is a useful method but is not always successful, especially for very rare alleles or if the original GWAS chip does not provide a robust scaffold for the imputation algorithm.

The 1KG Project has sequenced the whole genomes of a large number of individuals, but this still does not mean that we have a complete list of all SNPs in the human genome. Due to chance or population specificity, some very rare SNPs remain unknown, while some variants detected in this project may just be private variants unique to those individuals and may not be seen in other individuals. If the existing data from HapMap or 1KG do not support causality for the tagging SNP or its ssSNPs (the analyses are outlined in the next section), it is possible that an as-yet unknown SNP in LD with the tagging SNP may be responsible for the association signal. This type of failure is increasingly unlikely, but in the case that no known SNP in the vicinity of the association signal appears to be functional enough to be the causal SNP, the first thing to do is to explore whether there are any addi-tional whole-genome sequencing data from the population of interest. There are, for exam-ple, data sets from Icelandic, Dutch, and Asian Malay populations that are not included in the 1KG Project and which may include novel SNPs. Also, the African Genome Project pro-vides data from the genetically most diverse human populations. Most such data are acces-sible via major portals such as the NIH Database of Genotypes and Phenotypes (dbGaP) or the European Genome-phenome Archive (EGA). Analysis of whole-genome sequence data, however, requires assistance from bioinformaticians. Ultimately, it may be necessary to do additional sequencing in the region that provided the association signal. This is the basis of the **exome sequencing studies** that have been gaining popularity.

There is another bioinformatics approach for checking whether an unknown SNP may have caused the association signal. An ambitious project called Combined Annotation Depen-dent Depletion (CADD) has functionally assessed each and every possible nucleotide sub-stitution in the human reference sequence and generated a functionality score for the almost nine billion actual or potential SNPs (more details of this approach are given in the next section). The CADD data can be downloaded and inspected for the presence of highly func-tional nucleotide substitutions in the area where the association signal is located. If there is a strong suggestion of a functional variant, a post-GWAS candidate gene study on a sample from diseased individuals may be used to examine the presence or absence of such a variant. Of course, if the causal SNP is not in the immediate flanking region of the trait-associated SNP, the CADD analysis will not be useful. If all dry laboratory approaches fail, an additional sequencing study is the only option left to identify candidates for causal SNPs.

10.2 Functional Annotation of SNPs

SNPs contribute to the causation of common diseases by altering processes going on in the genome. Assigning roles to a SNP is called functional annotation. The most common effects are an altered expression level of a target gene, changes to amino acid sequence, or splicing alterations of the host gene (**Table 10.2**). These and other possible effects can either be computationally predicted or examined in existing data. Neither of these processes can rule out a SNP as nonfunctional, but if there is a strong indication of functionality, supported by empirical data from the appropriate tissue relevant to the disease process, the SNP may be considered to have a greater potential to be the causal SNP than other members of the ssSNP set. The examples below illustrate the power of bioinformatics analysis.

Examination of rs1507274 for functionality

dbGaP shows a number of associations with osteoporosis, one of which is rs1507274 (T > A). This is an intergenic SNP on Chromosome 1 located near *COL24A1*. Since it is an intergenic SNP, it is either a tagging SNP or, if causal, likely to have a regulatory role. As will be presented later in this chapter, RegulomeDB is a good tool to use to examine the functionality of SNPs with a potential regulatory role. Analysis with RegulomeDB reveals that rs1507274 matches the binding-site sequence (motif) for several transcription factors, including the human transcription factor ZNF410 (also known as APA-1) (**Figure 10.3**). A single change in DNA sequence can affect the binding motifs of multiple transcription factors, as in this example. Having the A allele of rs1507274 prevents binding of transcription factor ZNF410, but in order to attribute a role to rs1507274 as a causal SNP in the development of osteoporosis, ZNF410 must be involved in bone development or remodeling. A literature search on PubMed reveals a publication reporting that ZNF410 induces extracellular matrix-remodeling genes in fibroblasts, including bone marrow stromal fibroblasts (PubMed ID, PMID: 12370286). These results suggest a causal role for this SNP in osteoporosis development, but ultimate functional validation requires experimental verification. Besides, the ssSNP set may contain SNPs with stronger effects.

Examination of SNP rs610932 for functionality

SNP rs610932 is one of the aging-associated SNPs in the GWAS catalog and is located in the 3′ UTR of *MS4A6A*. RegulomeDB has no results on this SNP's functionality, it is not in

Table 10.2 Effects of different categories of SNPs

SNP location	SNP category	Affected gene
Intron Intergenic region Promoter 3′ UTR	Regulatory SNP	Target gene, whose expression is influenced (not necessarily the nearest gene; may be a gene on a different chromosome)
Exon	Coding-region SNP	Host gene (and any other gene if in a regulatory element)
Intron–exon junction	Splicing-site SNP	Host gene (and any other gene if in a regulatory element)

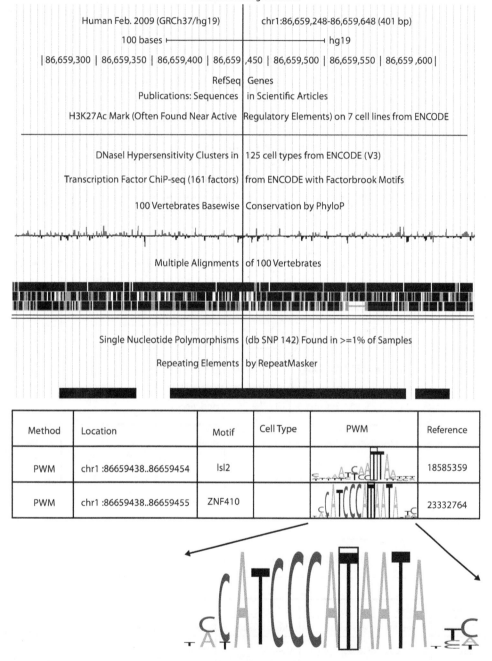

Figure 10.3 RegulomeDB analysis results for rs1507274. There are no experimental data to support the SNP being in a histone mark site or DNase hypersensitivity site, and no transcription factor binding to the SNP site has been detected. In the Motifs section of the report (lower panel), the position-weight matrix (PWM) for ZNF410 includes the SNP site (one of the dominant nucleotides, T, is changed to A by the SNP). ZNF140 is not one of the 161 transcription factors experimentally examined in the ENCODE project, but its motif contains the nucleotide substituted by the SNP.

a microRNA or transcription factor binding site, and it does not act as an eQTL. It is within a weak enhancer and does not change any binding motif for human transcription factors. A HaploReg search yields around 100 ssSNPs correlated with rs610932 with very high correlation coefficients ($r^2 \geq 0.80$). The nearest of those to rs610932, which provided the original association signal, is 2328 bp away: rs667897 (A > G; $r^2 = 0.87$). SNP rs667897 has the highest RegulomeDB functionality score among all the ssSNPs. It is located within a strong enhancer and within DNase hypersensitivity sites marking active transcription in 38 different cell types; it changes the sequence of binding motifs for three transcription factors, and acts as an eQTL for the nearby gene *MS4A4A* (**Figure 10.4**). Consequently, rs667897 has a very high RegulomeDB score and is very likely functional. One of the altered motifs is for transcription factor NFE2L2 (also known as NRF2). NFE2L2 is reported in the literature as a regulator of genes that contain antioxidant response elements in their promoters and encode proteins involved in the response to injury and inflammation, including the production of free radicals. Searching existing genetic association results on another online tool (GRASP) shows that rs667897 is already associated with late-onset Alzheimer

Motifs					
Method	Location	Motif	Cell Type	PWM	Reference
Footprinting	chr11:59936963..59936985	TCF11:MafG	Huvec		21106904
Footprinting	chr11:59936963..59936985	TCF11:MafG	Hpde6e6e7		21106904
Footprinting	chr11:59936963..59936985	TCF11:MafG	Helas3lfna4h		21106904
Footprinting	chr11:59936969..59936980	NFE2L2	Hpde6e6e7		21106904
Footprinting	chr11:59936969..59936980	NFE2L2	Huvec		21106904
PWM	chr11:59936969..59936980	NFE2L2			18006571
PWM	chr11:59936963..59936985	TCF11:MafG			16381825
PWM	chr11:59936968..59936981	FOXB1			23332764

Single nucleotides					
Method	Location	Affected Gene	Cell Type	Additional Info	Reference
eQTL	chr11:59936978..59936979	MS4A4A	Monocytes	cis	20502693

Figure 10.4 RegulomeDB analysis results for rs667897. In the Motifs section of the report (top panel), the position-weight matrix (PWM) for NFE2L2 includes the SNP site (one of the dominant nucleotides, A, is changed to G by the SNP). NFE2L2 is involved in the regulation of genes relevant to aging. Additionally, rs667897 is also an eQTL for *MS4A4A*, which is involved in Alzheimer disease development.

disease risk. The target gene for its eQTL effect, *MS4A4A*, is also known to be involved in late-onset Alzheimer disease development (PMID: 21460841). As a very likely functional SNP, rs667897 is possibly the causal SNP responsible for the association of rs610932 with aging, perhaps acting via more than one mechanism. This example illustrates how a simple bioinformatics analysis can suggest a mechanism for a known association that appears to be with a nonfunctional proxy SNP for the causal one. The generated hypothesis must then be tested experimentally for a conclusive result.

Descriptive information on SNPs can be obtained from genome browsers

Any work on the functional annotation of a SNP begins with gaining insight into its main features, such as its location on the chromosome (in a gene or between genes) or within a gene, and the genomic context or the characteristics of the region in which it is located (such as an evolutionarily conserved region, a **CpG island**, an **LD block**, or a region with histone modifications or a DNase hypersensitivity site). It is also important to have information on its frequencies in different populations. There are a number of genome browsers that compile this type of information for display on screen or for download. The most commonly used browsers to obtain descriptive information on genetic variation are the University of California Santa Cruz (UCSC), Ensembl, 1KG, and HapMap browsers. An example of the UCSC browser is given in **Figure 10.5**.

The UCSC browser can be queried by SNP, gene name, or genomic region. Different browsers use different formats for entering genomic regions. The human genome is constantly updated and the coordinates for genes and regions vary depending on the genome version or assembly. The assembly 19 coordinates for the *HFE* gene on Chromosome 6 can be given as either chr6:26,087,509-26,095,469 or chr6:26,087,509-26,095,469. The UCSC browser uses the first one. Entering these coordinates yields the result shown in Figure 10.5. Such coordinates can also be used if a region flanking the gene or multiple adjacent genes are of interest. Users also need to be careful that the genome assembly version they are using for genome coordinates matches the genome assembly version used by the genome browser. The UCSC browser clearly states which version it uses on the opening page, but other browsers may be ambiguous.

The UCSC results are presented in a long page that the user needs to scroll down to see all of the information. The top section of the results (Figure 10.5A) shows any gene that may be present at these coordinates with its exon–intron structure and all alternatively spliced forms. Any SNPs in the region that feature in the GWAS catalog are also shown. The *HFE* gene has two SNPs that have shown disease or trait associations, corresponding to H63D and C282Y mutations. Further down the results page (top of Figure 10.5B) are shown mRNAs and the location they correspond to in the reference sequence (usually the exons of any gene at this position), expressed sequence tags (ESTs) mapping to this region (which also usually correspond to exons), and CpG islands (the *HFE* region has none). The next result shown (Figure 10.5B) is from the ENCODE Project: the H3K27Ac histone modification that characterizes transcriptionally active parts of the genome—in this case, the beginning of the *HFE* gene. In genes with an alternative transcription start site, there may be another histone mark around the alternative site and any SNP in that area would be potentially functional. DNase hypersensitivity sites indicate transcriptionally active areas and usually coincide with transcription factor binding sites, which are the next feature shown in the results. Transcription factors are shown on the left of the screen if they have been found bound to the areas shown in the ENCODE Project experiments. Thus, these are empirical results and not computer predictions based on the presence of binding-site motifs. Any sequence variation in or around these sites could potentially abolish transcription factor binding. The results screen continues (Figure 10.5C) with chromatin and additional histone modifications (from the Broad Institute), which have different implications depending on the modification type; these data are followed by the DNA methylation and CpG methylation results from multiple cell types. The bottom of the long results screen (Figure 10.5D) shows further histone modification results, this time from different laboratories (Stanford, Yale, USC, and Harvard). Just below these are results for conserved transcription factor binding sites based on nucleotide sequence. This information complements the experimental ENCODE results seen in Figure 10.5B. These computed predictions for changes in transcription factor binding sites are useful because not all transcription factors have been examined experimentally. The display ends with SNPs in this region shown according to their coordinates. **Figure 10.6** shows how the number of categories (called features by the browser) can be reduced to the most relevant ones. In this example, only the gene structure and SNPs have been selected to be displayed after choosing to hide all the other categories one by one (either by right clicking on the category name on the screen or from the menu at the bottom of the page).

The UCSC browser is a commonly used online resource and a very comprehensive one. The Ensembl browser is similar to the UCSC browser. While the focus of the UCSC browser is on gene regulation, Ensembl's focus is more on genetic variation, sequence, and comparative genetics. One useful feature of Ensembl is the Variation Table (on the main menu) that produces a table of variants classified by their features (**Figure 10.7**). This table can be downloaded with details of each variant. If the number of variants is large, a link is

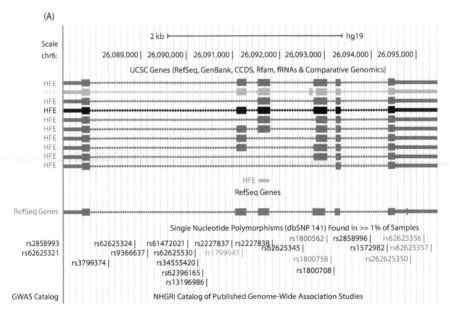

Figure 10.5 A typical display in the UCSC genome browser. In the gateway page, the gene name *HFE* was entered and these results were displayed for the corresponding genomic coordinates. The UCSC results are presented in a long page that is shown here divided into four parts (A–D) for ease of display; see the main text for details of the categories displayed. The menu that appears at the bottom of the on-screen page is not shown.

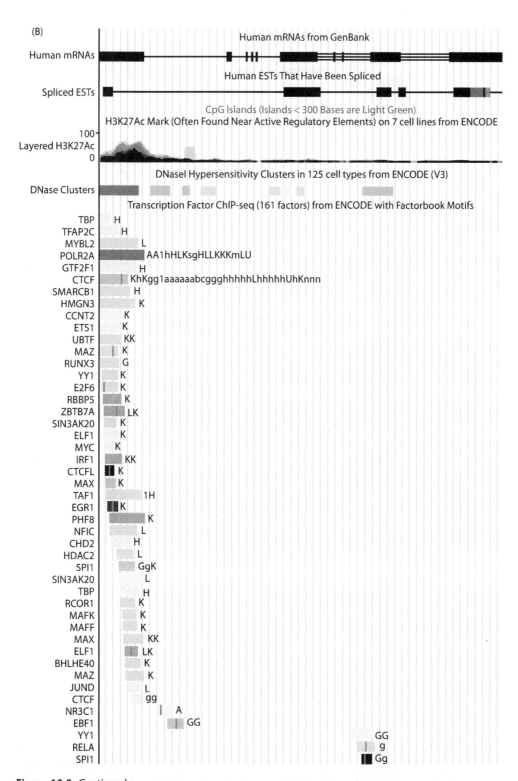

Figure 10.5 *Continued.*

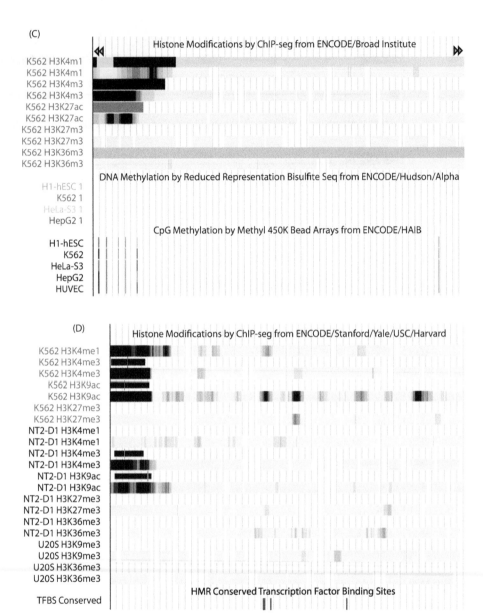

Figure 10.5 *Continued.*

provided to download it directly from the **BioMart facility**. The 1000 Genomes Project also uses Ensembl to display their results. The choice of the browser to use depends on the research question, but is largely a personal preference.

Individual SNPs can be assessed for different effects using online bioinformatics tools

There are a number of effects by which functional SNPs can contribute to the development of disease and there are dedicated bioinformatics tools to predict those effects. Many of these tools investigate the deleteriousness of amino acid substitutions caused by

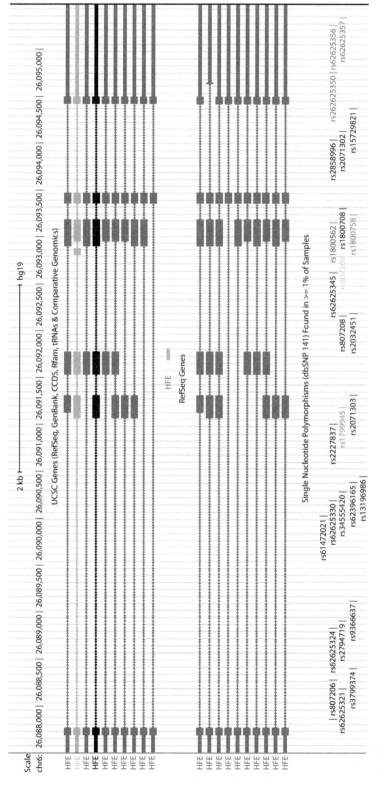

Figure 10.6 An example of displaying the selected features only by removing other features that are displayed by default. In this display from the UCSC browser, only the gene and SNPs are retained in order to inspect the locations of SNPs in the *HFE* gene.

Number of variant consequences	Type	Description
0	Transcript ablation	A feature ablation whereby the deleted region includes a transcript feature
22	Splice donor variant	A splice variant that changes the 2 base region at the 5' end of an intron
0	Splice acceptor variant	A splice variant that changes the 2 base region at the 3' end of an intron
43	Stop gained	A sequence variant whereby at least one base of a codon is changed, resulting in a premature stop codon, leading to a shortened transcript
12	Frameshift variant	A sequence variant which causes a disruption of the translational reading frame, because the number of nucleotides inserted or deleted is not a multiple of three
0	Stop lost	A sequence variant where at least one base of the terminator codon (stop) is changed, resulting in an elongated transcript
22	Initiator codon variant	A codon variant that changes at least one base of the first codon of the transcript
0	Transcript amplification	A feature amplification of a region containing a transcript
0	Inframe insertion	An inframe non synonymous variant that inserts bases into the coding sequence
18	Inframe deletion	An inframe non synonymous variant that deletes bases from the coding sequence
872	Missense variant	A sequence variant that changes one or more bases, resulting in a different amino acid sequence but where the length is preserved
66	Splice region variant	A sequence variant in which a change has occurred within the region of the splice site, either within 1-3 bases of the exon or 3-8 bases of the intron
0	Incomplete terminal codon variant	A sequence variant where at least one base of the final codon of an incompletely annotated transcript is changed
222	Synonymous variant	A sequence variant where there is no resulting change to the encoded amino acid
0	Stop retained variant	A sequence variant where at least one base in the terminator codon is changed, but the terminator remains
277	Coding sequence variant	A sequence variant that changes the coding sequence
0	Mature miRNA variant	A transcript variant located with the sequence of the mature miRNA
68	5 prime UTR variant	A UTR variant of the 5' UTR
164	3 prime UTR variant	A UTR variant of the 3' UTR
303	Non coding transcript exon variant	A sequence variant that changed non-coding exon sequence in a non-coding transcript
1977	Intron variant	A transcript variant occurring within an intron
0	NMD transcript variant	A variant in a transcript that is the target of NMD
539	Non coding transcript variant	A transcript variant of a non coding RNA gene
1411	Upstream gene variant	A sequence variant located 5' of a gene
1413	Downstream gene variant	A sequence variant located 3' of a gene
0	Protein altering variant	A sequence variant which is predicted to change the protein encoded within the coding sequence
6824	ALL	All variations

Figure 10.7 The Variation Table for the *HFE* gene on the Ensembl browser. The Ensembl gene ID is ENSG00000010704. The table shows the number of variants in each category plus a description, and provides links to show and download each category in detail. The whole table can be downloaded directly or, depending on its size, via the BioMart facility.

missense SNPs. The discovery that more than 80% of GWAS top hits are in noncoding regions of the genome shifted the emphasis toward **regulatory SNPs** (rSNPs) that modify gene expression. However, the advent of exome sequencing studies as a popular post-GWAS approach has renewed interest in missense SNPs. Selected bioinformatics resources for the assessment of the effects of individual SNPs are listed in **Table 10.3**.

Table 10.3 Bioinformatics resources for the assessment of individual effects of SNPs

Resource	Type of variant/input	Function tested	Features
PolyPhen2 (Polymorphism Phenotyping)	Missense/rs ID	Deleterious effect on protein function	Based on sequence conservation in a multiple sequence alignment, structure-to-model position of amino acid substitution, and SWISS-PROT annotation
SIFT (Sorting Intolerant From Tolerant)	Missense/rs ID	Deleterious effect on protein function	Based on sequence conservation
SNPs3D	Missense/rs ID	Deleterious effect on protein function	Based on structure- and sequence-based support vector machines Provides a score for the effect of the SNP on protein function. Also includes gene and gene interactions based on text mining
MutPred	Protein sequence variation	Classifies the protein mutation as disease-causing or neutral	Requires a protein sequence in FASTA format and a list of amino acid substitutions for analysis
dbSTEP (Database of Splice Translational Efficiency Polymorphisms)	Splice translational efficiency polymorphisms (STEPs) that alter splicing of the 5′ UTR of pre-mRNAs (potentially affecting protein quantity but not quality)	Effect on protein quantity	Allows searching by gene, SNP, or genomic coordinates and provides a detailed report. It cross-matches to known SNP associations in GWASs
AASsites (Automatic Analysis of SNP Sites)	The input is the DNA sequence; the program identifies the SNP	Effect on protein sequence (splicing pattern)	Identifies the splice regulatory site and predicts its effect on protein function
GTEx (Gene and Tissue Expression)	Any SNP/rs ID	Effect on tissue-specific gene expression level	Currently the most comprehensive database for eQTLs and with tissue specificity; uses an extreme statistical significance threshold (no eQTL for most genes)
SCAN	Any SNP/rs ID	Effect on tissue-specific gene expression level	Samples are from the original HapMap populations and EBV-transformed lymphoblastoid cell lines
eQTL Browser (UChicago)	Any SNP/rs ID (or gene or genomic region)	Effect on tissue-specific gene expression level	Integrates results from more than 15 studies into a genome browser
Blood eQTL Browser	Any SNP/rs ID or gene	Effect on gene expression level in blood cells	Results of a meta-analysis study

Web addresses are provided in the URL list at the end of the chapter.

The algorithms PolyPhen2 and SIFT are the most established resources for missense SNP assessment. Both algorithms use a 0-to-1 scale, but PolyPhen2 considers the SNPs with the greater score as damaging, while SIFT considers them as tolerable substitutions. While PolyPhen2 and SIFT only provide results for existing missense SNPs, potentially any amino acid can mutate to another one, even though no example has yet been noted. The PMut algorithm provides precalculated scores for all the positions (19 possible mutants per position) for all proteins, although only a fraction of them have been detected.

Large numbers of SNPs can be simultaneously assessed for multiple effects using integrative analysis tools

Most of the time, it is not a single SNP but a set of SNPs that needs to be analyzed for functionality. The tools listed in Table 10.3 can be used for multiple SNP analysis, but it is a time-consuming and tedious process. Fortunately, there are resources that allow simultaneous annotation of multiple SNPs for multiple functions that compile results from most of the resources listed in Table 10.3. **Table 10.4** lists some of the commonly used tools that aggregate annotations from different resources.

Actual and potential SNPs already have functionality scores

When a group of SNPs is assessed for their effects on gene function, it would be useful to be able to rank them from the most deleterious to least deleterious, and many algorithms can integrate information from multiple resources to generate a single score to reflect a SNP's effect. The most recently introduced and most comprehensive of these scores are RegulomeDB, CADD, and DANN scores.

RegulomeDB is a curated database of noncoding SNPs that can be queried for a single SNP or multiple SNPs as well as for genomic regions. Representative results are shown in Figures 10.3 and 10.4. The RegulomeDB Web server integrates mainly experimental ENCODE data and computer predictions on the potential gene regulatory function of noncoding-region SNPs. The results are presented as a RegulomeDB score, the highest being 1a and the lowest being 6. The higher the score, the greater are the effects of the SNP on gene function. SNPs scoring 1a would have the highest functionality. A score of 7 is given if there are no available data to assess, and unfortunately, the proportion of SNPs with this score may be high, depending on the genomic region. A full assessment page for a SNP provides all the details for the background to the score given, including which transcription factor binds to the SNP site in which tissue, whether there is a DNase hypersensitivity site near the SNP, or if any protein has been found bound to the region in experiments and in which tissue. RegulomeDB also offers lists of pre-analyzed GWAS results, with scores for lead SNPs that are reported in the GWAS along with scores for their ssSNP sets.

As RegulomeDB focuses on noncoding SNPs, a missense SNP may not score highly unless it also happens to be in a regulatory region. For example, the disease-causing *HFE* mutation C282Y (rs1800562) scores 5, corresponding to minimal evidence for functionality, regardless of its effect on protein function. For a fully integrative functionality score, the Combined Annotation Dependent Depletion (CADD) algorithm has been developed, although specialized scores by other algorithms for missense SNPs may be more accurate (for missense SNPs). The CADD score is presented in two forms: the raw and scaled C-score. The CADD algorithm uses the Ensembl Variant Effect Predictor, data from the

Table 10.4 Bioinformatics resources for integrative assessment of SNP effects

Resource	Input	Output	Features
PheGenI (Phenotype-Genotype Integrator)	SNP (rs ID), gene, or genomic region (up to around 400 SNPs at a time); allows restriction to a SNP category (exonic, intronic, UTR, or near gene)	Descriptive data on SNPs, the genes they map to, eQTL data (from NCBI eQTL browser), and association results from GWAS catalog and dbGaP Association Results Browser	Annotation of any SNP; most useful to obtain eQTL and association results in one stop; each section of the results is downloadable separately
SNPnexus	Up to 100,000 SNPs (rs ID) or 1 Mb genomic region	On-screen display; downloadable as a single spreadsheet with multiple worksheets	Annotation of any SNP; reports many possible functional consequences from the major gene annotation systems (including PolyPhen, SIFT, microRNA, or microRNA binding site; transcription factor binding site; CpG site, GWAS catalog, and COSMIC database for cancer somatic mutations)
Ensembl Variant Effect Predictor	Multiple options (from single rs IDs to a text file with multiple entries)	On screen; can be downloaded	A Web-based service with Perl scripts and application programming interface (API); annotation of any SNP including unknown SNPs (by position and nucleotide substitution)
AVIA (Annotation, Visualization and Impact Analysis)	SNP information in choice of formats or a text file	E-mailed; very comprehensive	AVIA is a Web server that annotates many aspects of the variants, including functional consequences; incorporates multiple scoring algorithms (scoring algorithms for missense SNPs include PolyPhen2, SIFT, VEST, CADD, Mutation Taster, and Mutation Assessor). AVIA v2.0 extends its original implementation by adding information focusing on epigenetics, gene expression, and protein annotations
SNiPA (Single Nucleotide Polymorphisms Annotator)	SNP (rs ID), gene symbol, or position	Recombination rate and functional annotation	A Web service offering variant-centered genome browsing and interactive visualization tools tailored for easy inspection of many variants in their locus context. Also provides the CADD score and eQTL targets list
CRAVAT (Cancer-Related Analysis of Variants Toolkit)	Web entry or text file in a simple format (chromosome, position, strand, reference, and alternative nucleotides)	E-mailed when completed as spreadsheet files (text files optional)	Analyzes any variant (does not have to be a cancer-related one); incorporates Variant Effect Scoring Tool (VEST) analysis; most comprehensive output as text or spreadsheet

(Continued)

Table 10.4 Bioinformatics resources for integrative assessment of SNP effects (*Continued*)

Resource	Input	Output	Features
ANNOVAR and **wANNOVAR** (ANNOtate VARiation)	A variant call format (VCF) text file or online entry	Stand-alone desktop version produces results in files; the Web interface version sends an e-mail when downloadable results are ready	A large suite of variant annotation tools, which can also be used online via the Web interface (wANNOVAR). Primarily designed for next-generation sequencing studies, but works with small lists of variants as well. The ANNOVAR server keeps on updating the existing resources it uses
dbNSFP (Database for NonSynonymous SNPs' Functional Predictions)	No input necessary	A database of annotations of all potential nonsynonymous and splice-site SNPs in the human genome	The dbNSFP is an integrated database of 14 functional prediction scores (SIFT, PolyPhen2, LRT, MutationTaster, MutationAssessor, FATHMM, VEST3, CADD, MetaLR, MetaSVM, PROVEAN, DANN, fathmm-MKL, and fitCons), 6 conservation scores (PhyloP $\times$ 2, phastCons $\times$ 2, GERP++, and SiPhy), and other related information including allele frequencies
SnpEff	A VCF text file	It annotates the variants and calculates the effects they produce on known genes	SnpEff is a genetic variant annotation and effect prediction toolbox. Can also analyze variants in non-human genomes. There is no Web interface and the stand-alone software is for more advanced users

Web addresses are provided in the URL list at the end of the chapter.

ENCODE project, and information already integrated within the UCSC genome browser. The integration spans a wide range of data types including conservation metrics, functional genomic data such as DNase hypersensitivity and transcription factor binding, and transcript information such as distance to splice sites or expression levels in cell lines. Protein-level scores such as PolyPhen2 , SIFT, and Grantham are also used. The scaled C-scores range from 1 to 99, based on the rank of each variant relative to all possible 8.6 billion substitutions in the human reference genome. The higher the scaled C-score for a SNP, the stronger is the evidence for its functionality. A scaled C-score $\geq$10 indicates a SNP predicted to be among the top 10% of the most deleterious SNPs in the human genome. CADD scores can be obtained by submitting a list of SNPs in a special format to the CADD Web server. All CADD scores for the 8.6 billion actual and potential SNPs in the human genome are available for download, but the file size is prohibitively large for nonexpert users. The linear kernel support vector machine-based algorithm used in CADD analysis has been improved by using a deep neural network (DNN), which also considers nonlinear effects. This algorithm is called Deleterious Annotation of genetic variants using Neural Networks (DANN) and also provides a score. Currently, the latest scores are only available for download as a whole set and have not yet been independently and fully assessed for accuracy.

10.3 Analyzing Functionally Altered Genes

SNPs are associated with diseases as a result of the SNP altering gene function. Since multiple SNPs are involved in complex disease pathogenesis, and each SNP also has a ssSNP set, the cumulative effect may be many genes being functionally altered. The identity of these genes is a useful insight into disease pathophysiology, but even more information may be gained if the affected gene list is subject to additional analyses as a set.

A SNP list can be converted to a gene list for further analysis

Although the most common effect of a disease-associated SNP is alteration of gene expression levels, other effects are possible, as listed in Table 10.2. SNPs in certain locations may affect multiple genes: a microRNA or transcription factor gene sequence, a regulatory region, or a splice site. Possession of the minor allele of a SNP means possession of minor alleles of all other SNPs in absolute LD ($r^2 = 1.0$) with that SNP. The SNPs in absolute LD are known as perfect proxy SNPs and should be taken into account when assembling a gene list. It is even justifiable to relax the LD threshold for proxy SNP selection, since causal SNPs may be in moderate LD ($r^2 \sim 0.50$) with the tagging SNPs associated with the trait. Once the SNP list is complete, each SNP should be subjected to individual annotation to identify the affected genes, which may be either the host gene or a different target (see Table 10.2).

As an example, the *HFE* SNP rs1800562 is a missense SNP, and the substitution of an amino acid at position 282 (C282Y) causes a drastic change in the HFE protein; this results in a very high CADD score of 26.4, indicating high deleteriousness. Thus, the *HFE* gene is directly affected by rs1800562. However, rs1800562 also shows eQTL effects on *HFE*, *ALAS2*, and a few other genes (SNiPA summarizes these results), and these genes should be included in the target gene list. HaploReg analysis shows that there is one SNP in absolute LD with rs1800562 (rs79220007) and two more SNPs in strong LD ($r^2 > 0.80$). The genes affected by these SNPs should be identified by functional annotation of the additional SNPs, and a complete list of genes should be prepared for further analysis.

A gene list can be analyzed for enrichment

All the genes in the compiled list can be analyzed for any common features that they share. The most commonly examined features include participation in a common pathway, sharing of the same gene ontology annotation (essentially having the same function), regulation by the same transcription factor or microRNA, or having been identified in the same gene sets in transcriptomics studies (such as being up-regulated by the same drug). This examination has been automated by various online resources (**Table 10.5**). Any common features found are statistically assessed for significance. For example, if a gene list contains 150 genes, and 30 of them belong to the WNT signaling pathway, then 20% of the genes in the list are in the WNT pathway. Among the 20,000 genes in the human genome, if 200 of them belong to the same pathway, then 1% of all genes are in the same pathway. These two proportions are then statistically compared and if the difference is statistically significant, it is concluded that genes of the WNT signaling pathway are disproportionately affected (by the SNPs in the study). This is usually phrased as the gene set being enriched in genes involved in the WNT signaling pathway. Enrichment can be sought for many different features, and this is done by converting a SNP list to a gene list and submitting the gene list to one of the online resources listed in Table 10.5. There are also commercially available software packages that can do enrichment analysis starting from the SNP list.

Table 10.5 Selected examples of online resources for enrichment analysis

Resource	Input	Output	Features
DAVID (Database for Annotation, Visualization and Integrated Discovery)	A gene list (a gene identifier converter is provided)	On-screen display	The prototype pathway analysis tool using a gene list. A full protocol paper is included in the Further Reading list. Also lists interacting proteins
GSEA/MSigDB (Gene Set Enrichment Analysis/ Molecular Signatures Database)	Gene list	On-screen display	This resource examines a gene list for enrichment in pathways as well as in molecular signature gene sets (including oncogenic and immunologic signatures, positional and gene ontology gene sets, and motif gene sets)
WebGestalt (WEB-based GEne SeT AnaLysis Toolkit)	Gene list (multiple identifiers, including microarray gene IDs, are accepted)	On-screen display	One of the most comprehensive functional enrichment tools, which is frequently updated (and can examine gene lists from multiple organisms)
ToppFun (Transcriptome, ontology, phenotype, proteome, and pharmacome annotations-based gene list functional enrichment analysis)	Gene list	On-screen display	Examines enrichment of gene lists based on transcriptome, proteome, regulome (transcription factor binding sites and miRNA), gene ontology and pathways, pharmacome (drug–gene associations), and literature co-citation
UniHI (Unified Human Interactome)	Gene or protein lists	On-screen display	Besides functional enrichment analysis, provides detailed information on protein interactions
HumanMine (integrated genomics, genetics and proteomics data warehouse)	Gene, protein, or SNP lists	On-screen display	This tool integrates data from genomics and proteomics domains. Besides gene lists, protein and SNP lists are also analyzed. Focuses on metabolic disorders

The resources presented in this chapter are powerful tools for the functional annotation of SNPs. Nonspecialist users of online resources are limited by the options given by the Web server and limits on the size of the queries. Anyone seriously intent on running genome-wide bioinformatics analysis without restrictions is recommended to become familiar with the R program and the accompanying package Bioconductor, which is designed specifically for bioinformatics analysis.

Key Points

- In its early days, bioinformatics was mainly about computational prediction of functional elements in DNA sequences. Contemporary bioinformatics is more about the integration of large sets of biological data and computer-based assessment to inform genome biology.

- Most GWAS results are indirect associations and provide an association signal indicating the presence of a causal SNP in the vicinity. Post-GWAS approaches explore the area for the exact source of the signal.

- Bioinformatics analysis integrates genomics, transcriptomics, methylomics, proteomics, and metabolomics results to identify functional SNPs within the statistically similar SNP set as the source of the association signal.

- With the completion of comprehensive projects like ENCODE and Epigenomics Roadmap, data from many cell and tissue types are available and can be interrogated simultaneously using bioinformatics. The availability of such data represents an advantage over wet laboratory experiments, which are not easy to carry out on multiple cell types.

- The most prominent effect of a SNP in disease causation is on gene expression regulation. RegulomeDB provides functionality scores for regulatory SNPs.

- There are databases for the precalculated functionality scores of billions of actual and theoretical SNPs (for example, CADD and DANN) and protein mutations (PMut) and these can be downloaded or interrogated via Web servers. The availability of this information facilitates the ranking of the multiple variants under consideration for causality assessment.

- Confidence levels in bioinformatics results can be very high, but it is advisable to check more than one algorithm, and all results require experimental verification using the specific cell or tissue type.

URL List

This list is a selected subset of many resources available. For the most up-to-date and complete list of bioinformatics tools, see http://www.dorak.info/mtd/bioinf.html

1000 Genomes Browser. http://browser.1000genomes.org

AASsites (Automatic Analysis of SNP Sites). http://genius.embnet.dkfz-heidelberg.de/menu/cgi-bin/w2h-open/w2h.open/w2h.startthis?SIMGO=w2h%2ewelcome

ANNOVAR (ANNOtate VARiation). http://www.openbioinformatics.org/annovar

AVIA (Annotation, Visualization and Impact Analysis). http://avia.abcc.ncifcrf.gov/apps/site/sub_analysis/?id=3 (See also the chart comparing AVIA to other tools at http://avia.abcc.ncifcrf.gov/apps/site/compare)

Bioconductor. http://www.bioconductor.org

Blood eQTL browser. http://www.genenetwork.nl/bloodeqtlbrowser

CADD (Combined Annotation Dependent Depletion). University of Washington. http://cadd.gs.washington.edu

CRAVAT (Cancer-Related Analysis of Variants Toolkit). Institute for Computational Medicine, The Johns Hopkins University. http://www.cravat.us

DANN (Deleterious Annotation of genetic variants using Neural Networks). https://cbcl.ics.uci.edu/public_data/DANN/data

DAVID (Database for Annotation, Visualization and Integrated Discovery). http://david.abcc.ncifcrf.gov

dbGaP (Database of Genotypes and Phenotypes). National Center for Biotechnology Information. http://www.ncbi.nlm.nih.gov/gap

dbGaP Association Results Browser. National Center for Biotechnology Information. http://www.ncbi.nlm.nih.gov/projects/gapplusprev/sgap_plus.htm

dbNSFP (Database for NonSynonymous SNP's Functional Predictions). Jpopgen. https://sites.google.com/site/jpopgen/dbNSFP

dbSTEP (Database of Splice Translational Efficiency Polymorphisms). http://epi-sharp.epi.bris.ac.uk/dbstep/dbstep.cgi

EGA (European Genome-phenome Archive). EMBL-EBI. https://www.ebi.ac.uk/ega

Ensembl. Human genome browser. Wellcome Trust Sanger Institute/EMBL-EBI. http://www.ensembl.org/Homo_sapiens

Ensembl Variant Effect Predictor. Wellcome Trust Sanger Institute/EMBL-EBI. http://www.ensembl.org/info/docs/tools/vep

ENTREZ SNP (dbSNP). National Center for Biotechnology Information. http://www.ncbi.nlm.nih.gov/snp

eQTL Browser (UChicago). http://eqtl.uchicago.edu/cgi-bin/gbrowse/eqtl

GenEpi Toolbox (A collection of bioinformatic tools for genetic epidemiology). http://genepi_toolbox.i-med.ac.at

GLIDERS Genome-wide LInkage DisEquilibrium Repository and Search Engine). Wellcome Trust Sanger Institute. https://www.sanger.ac.uk/resources/software/gliders

GRASP (Genome-wide Repository of Associations between SNPs and Phenotypes). National Heart, Lung, and Blood Institute. National Institutes of Health. http://apps.nhlbi.nih.gov/Grasp/Search.aspx

GSEA/MSigDB (Gene Set Enrichment Analysis). Broad Institute. http://www.broadinstitute.org/gsea/msigdb/annotate.jsp

GTEx (Gene and Tissue Expression) Portal. Broad Institute. http://www.gtexportal.org/home

HaploReg v3. Broad Institute. http://www.broadinstitute.org/mammals/haploreg/haploreg_v3.php

Haploview. Broad Institute. https://www.broadinstitute.org/scientific-community/science/programs/medical-and-population-genetics/haploview/haploview

HapMap Browser. International HapMap Project. http://hapmap.ncbi.nlm.nih.gov/cgi-perl/gbrowse/hapmap28_B36

HumanMine. Department of Genetics. University of Cambridge. http://www.humanmine.org

MutPred. Buck Institute. Indiana University. http://mutpred.mutdb.org

NHGRI GWAS Catalog (A Catalog of Published Genome-Wide Association Studies). National Human Genome Research Institute. National Institutes of Health. http://www.genome.gov/gwastudies

PheGenI (Phenotype-Genotype Integrator). National Center for Biotechnology Information. http://www.ncbi.nlm.nih.gov/gap/phegeni

PMut (Pathological Mutations). Molecular Modelling and Bioinformatics Group. Institute for Research in Biomedicine Barcelona. http://mmb.pcb.ub.es/PMut

PolyPhen2 (Prediction of functional effects of human nsSNPs). http://genetics.bwh.harvard.edu/pph2

R Project for Statistical Computing. The R Foundation. http://www.r-project.org

RegulomeDB. Center for Genomics and Personalized Medicine at Stanford University. http://www.regulomedb.org

RegulomeDB (Linking Disease Associations with Regulatory Information in the Human Genome). GWAS Companion website. Stanford University. http://regulome.stanford.edu/GWAS

rSNPBase (a database for curated regulatory SNPs). Bioinformatics Lab. Institute of Psychology. Chinese Academy of Sciences: http://rsnp.psych.ac.cn/listSearch.do

SCAN (SNP and CNV Annotation Database). University of Chicago. http://www.scandb.org

SIFT (Sorting Intolerant from Tolerant). http://blocks.fhcrc.org/sift/SIFT.html

SNAP (SNP Annotation and Proxy Search). Broad Institute. http://www.broadinstitute.org/mpg/snap

SnpEff (Genetic variant annotation and effect prediction toolbox). http://snpeff.sourceforge.net

SNPnexus. Barts Cancer Institute. http://www.snp-nexus.org

SNPs3D. http://www.snps3d.org

ssSNPer (web interface to identify statistically similar SNPs (ssSNPs) in the HapMap), http://neurogenetics.qimrberghofer.edu.au/ssSNPer

ToppFun (Transcriptome, ontology, phenotype, proteome, and pharmacome annotations based gene list functional enrichment analysis). ToppGene Suite. Cincinnati Children's Hospital Medical Center. https://toppgene.cchmc.org/enrichment.jsp

UCSC Genome Bioinformatics. University of California Santa Cruz. http://genome.ucsc.edu

UniHI (Unified Human Interactome). http://www.unihi.org

VaDE (VarySysDB Disease Edition). SNP Function. Biomedical Informatics Laboratory. Tokai University School of Medicine. http://bmi-tokai.jp/VaDE/snp-annotation

wANNOVAR (Web-based access to ANNOVAR software). http://wannovar.usc.edu

WebGestalt (WEB-based GEne SeT AnaLysis Toolkit). http://www.webgestalt.org

Further Reading

Arnold M, Raffler J, Pfeufer A et al. (2015) SNiPA: an interactive, genetic variant-centered annotation browser. *Bioinformatics* 31, 1334–1336 (doi: 10.1093/bioinformatics/btu779).
(*Explains the process involved in development of one of the latest integrative functional annotation tools. The supplementary data file lists all bioinformatics tools incorporated in SNiPA with their description, links, and references.*)

Auer PL & Lettre G (2015) Rare variant association studies: considerations, challenges and opportunities. *Genome Med* 7, 16 (doi: 10.1186/s13073-015-0138-2). (*With rare variant association studies becoming more common, bioinformatic analysis of rare variants is also becoming an important issue. This review highlights important aspects of functional annotation of rare variants, provides a list of available tools, and gives illustrated examples.*)

Bromberg Y (2013) Chapter 15: Disease gene prioritization. *PLoS Comput Biol* 9, e1002902 (doi: 10.1371/journal.pcbi.1002902). (*An overview of analysis based on gene lists, focusing on protein–protein interactions. It comes with a glossary, its own further reading list, and suggested exercises and their answers.*)

Coassin S, Brandstätter A & Kronenberg F (2010) Lost in the space of bioinformatic tools: a constantly updated survival guide for genetic epidemiology. The GenEpi Toolbox. *Atherosclerosis* 209, 321–335 (doi: 10.1016/j.atherosclerosis.2009.10.026). (*Introduces the GenEpi Toolbox and reviews the bioinformatics tools used by genetic epidemiologists for SNP selection and SNP annotation.*)

Huang da W, Sherman BT & Lempicki RA (2009) Systematic and integrative analysis of large gene lists using DAVID bioinformatics resources. *Nat Protoc* 4, 44–57 (doi: 10.1038/nprot.2008.211). (*This paper presents a complete protocol for analyzing a gene set by DAVID for enrichment analysis.*)

Karchin R (2009) Next generation tools for the annotation of human SNPs. *Brief Bioinform* 10, 35–52 (doi: 10.1093/bib/bbn047). (*Reviews most of the currently used tools with examples as case studies, and provides guidance in interpretation of results.*)

Knight JC (2014) Approaches for establishing the function of regulatory genetic variants involved in disease. *Genome Med* 6, 92 (doi: 10.1186/s13073-014-0092-4).

Lee PH & Shatkay H (2008) F-SNP: computationally predicted functional SNPs for disease association studies. *Nucleic Acids Res* 36(Database issue), D820–D824 (doi: 10.1093/nar/gkm904). (*Most useful for appreciation of the decision procedure for functional SNP assessment.*)

Li L & Wei D (2015) Bioinformatics tools for discovery and functional analysis of single nucleotide polymorphisms. *Adv Exp Med Biol* 827, 287–310 (doi: 10.1007/978-94-017-9245-5_17).

Mooney SD, Krishnan VG & Evani US (2010) Bioinformatic tools for identifying disease gene and SNP candidates. *Methods Mol Biol* 628, 307–319 (doi: 10.1007/978-1-60327-367-1_17). (*This can be seen as an expansion of Table 10.3, with further details and more tools for the assessment of each function.*)

Wang J, Duncan D, Shi Z & Zhang B (2013) WEB-based GEne SeT AnaLysis Toolkit (WebGestalt): update 2013. *Nucleic Acids Res* 41(Web Server issue):W77–W83 (doi: 10.1093/nar/gkt439). (*Explains the process of interpreting gene lists derived from large-scale genetic, transcriptomic, and proteomic studies in functional enrichment analysis, and introduces one of the enrichment analysis tools, WebGestalt.*)

Wang P, Dai M, Xuan W et al. (2006) SNP Function Portal: a web database for exploring the function implication of SNP alleles. *Bioinformatics* 22, e523–e529 (doi: 10.1093/bioinformatics/btl241). (*Although the tool introduced in this paper is now not up to date, the thought process described in this paper as a background to bioinformatic analysis of SNP functionality is very informative.*)

Ward LD & Kellis M (2012) Interpreting noncoding genetic variation in complex traits and human disease. *Nat Biotechnol* 30, 1095–1106 (doi: 10.1038/nbt.2422). (*This review by the creators of HaploReg covers a brief history of disease genetics, the diversity of genetic architecture underlying human diseases, the differences between coding- and noncoding-region variants with their functional annotations and mechanisms of their effects on genes, and tools for noncoding variant analysis.*)

Yang H & Wang K (2015) Genomic variant annotation and prioritization with ANNOVAR and wANNOVAR. *Nat Protoc* 10, 1556–1566 (doi: 10.1038/nprot.2015.105). (*Introduces the command-line tool ANNOVAR for expert users, the input file format (variant call format, VCF), and the Web interface version of ANNOVAR (wANNOVAR). A complete protocol for hands-on practice is provided.*)

Genetic Risk Profiling and Medical Applications of Genetic Associations

11

Genetic association study results serve three main purposes: (1) to identify risk markers that can be used to predict high-risk individuals who may benefit from earlier and more intensive interventions, as well as to define prognostic categories; (2) to gain insight into disease biology; and (3) to predict the response or degree of response to treatment (**Figure 11.1**). To date, the second aim has been the best fulfilled and the third aim has been of some clinical use. There has been no such success in translating the discovery of genetic markers to clinically useful markers for the prediction of future disease development and population screening. Given the number of statistically significant GWAS results, some with P values exceeding 10^{-250}, it may sound paradoxical that we do not yet have predictive markers for disease development. This chapter discusses why there is, in fact, no paradox and also contrasts the current status of genetic risk profiling with the biological information obtained from genetic association studies. The development of biomarkers and assessment of their clinical utility is also covered. The emphasis is on GWASs, as they provide more complete coverage of the genome for identification of risk markers, but genetic risk profiling is also possible using the results of candidate gene studies. The latter approach would obviously generate a more limited risk profile.

11.1 Genetic Risk Profiling and Genetic Markers

Genetic risk profiling aims to classify individuals, based on their genotypes, as high risk or low risk for future disease development. When GWASs started reporting statistically highly impressive results from genome-wide screens, it was thought that these results would lead to a new era of effective future risk prediction among healthy people. It was soon realized that this was not an easy task.

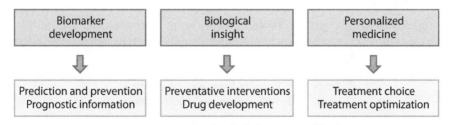

Figure 11.1 The aims of genetic association studies. The biomarker development aim has not been fulfilled yet, other than for pharmacogenetic markers, but several genetic markers are already in use for aiding treatment choice and optimization. The studies have offered much biological insight into disease development.

There are experimental, genetic, and statistical reasons why GWAS results have not resulted in effective risk profiling

Most GWAS results are obtained in case-control studies, but genetic risk profiling will be performed at the population level where susceptible people are a minority. A marker may be more frequent in diseased individuals compared with healthy ones, but possession of the marker alone does not determine the development of the disease. The initial, and perhaps greatest, success in disease genetics so far has been in **monogenic diseases** where **genetic determinism** operates: one mutation is necessary and sufficient to cause the disease, such as in cystic fibrosis or sickle-cell anemia. In complex disorders such as diabetes or schizophrenia, this is not the case. Even sets of strong genetic markers, when analyzed jointly, do not act deterministically for complex diseases despite yielding a strong effect size with statistical significance. With a good understanding of what is measured by statistical significance and what defines a good risk marker, it is easy to see that this outcome is logical.

In statistical significance testing of case-control studies, the magnitude of the difference in genotype frequencies between cases and controls is assessed: the larger the difference, the smaller the P value. This result comes up even when the marker is not exclusively present in the cases—that is, even when it has limited ability to classify individuals as cases or controls. The data in **Table 11.1** illustrate how statistical significance can be obtained even when the marker is present only in some cases and despite it not being able to classify individuals.

The results in **Table 11.1** indicate that the presence of the marker shows some correlation with the presence of the disease, but it is not an absolute correlation. The marker is never present in healthy controls (there are no false positives) but it is absent in a majority of the cases (false negatives). Statistical analysis of these results reveals a high statistical significance ($P < 0.0001$) and a more than fiftyfold increased risk (OR > 50) for individuals who possess the marker. Despite this impressive result, this marker is not a good classifier for individual risk profiling.

In the extreme example given above, the marker still identifies a small percentage of diseased subjects and is never present in healthy people, suggesting that it may have some use. At the population level, however, using this marker to identify high-risk individuals will require the testing of a very large number of subjects to detect one truly susceptible person, and a negative result will not rule out susceptibility. An ideal classifier labels those individuals with positive results as susceptible and all others as nonsusceptible. To judge how close a marker is to the ideal classifier status, additional statistical properties are taken into account.

The hypothetical example given in **Table 11.1** is extreme, in that the marker is present only in some cases and never in healthy people. This usually happens when researchers keep

Table 11.1 Results from a case-control study with 1000 cases and 1000 controls

	Cases	Controls
Marker-positive	50	0
Marker-negative	950	1000

combining markers until they are never present together in healthy people, but are present only in a fraction of cases. A more common scenario is that a genetic risk marker for a complex disease will always be present in a proportion of disease-free subjects. For example, the much-talked-about obesity risk marker *FTO* SNP rs9939609 variant allele is present in up to 45% of chromosomes in Europeans who are not obese.

Population screening amplifies small imperfections

Markers that initially appear to have perfect properties are not often useful for population screening. A marker that is present in 99% of patients with a rare disease and present only in 1% of healthy people appears to be a perfect classifier. When this marker is applied to a large number of healthy people to classify them for their disease risk, a different picture emerges. It may easily label more people at risk of developing the disease than the actual number of people truly at risk when screening the general population. For example, if one million people are screened for a very rare (1/100,000) disease, all ten subjects with the disease or at risk will be detected, but the 1% false-positivity rate will result in 10,000 subjects labeled as at risk for the disease when they are not. If the disease is more common (1/1000), 1000 out of one million tests will be true positives, but there will be 9990 false positives: ten times more false-positive results than true results. Thus, the frequency of the disease is important in the success of genetic risk profiling, and even small false-positivity rates mean that the marker will classify more people as at risk than those who actually are at risk (**Box 11.1**). Another important issue is the cost effectiveness, which is determined by the cost of each test, the ratio of true positives to false positives, and the number of tests necessary to detect one truly susceptible person. Due to unfavorable cost effectiveness, most genetic tests are currently used for differential diagnosis in the clinic rather than in population screening.

Box 11.1 Why do we not use *HFE* C282Y screening in the general population?

The C282Y mutation (rs1800562) of the *HFE* gene causes autosomal recessive hereditary hemochromatosis, an iron overload disease that manifests in later life. The disease causes multi-organ failure due to iron deposition and increases the risk for liver cancer up to 200 times. If diagnosed before the terminal effects occur, the treatment is simple and consists of blood withdrawals to remove excess iron from the body. The mutation is quite common in Western Europeans and up to 15% of chromosomes may carry the mutation. Genotyping for C282Y is straightforward and genotyping error rate is negligible.

When the mutation was first discovered, it was argued that population screening could be used to identify individuals with the mutation and the disease prevented by dietary intervention to reduce iron intake and absorption, as well as encouraging regular blood donations in mutation-positive individuals. Since every hereditary hemochromatosis patient is positive for the mutation, while a mutation-negative person very rarely develops the disease (when it is caused by another low penetrance mutation, H63D) there was initial enthusiasm about population screening. However it was soon realized that there were also a lot of mutation-positives who did not develop the disease. In fact, only 5% of mutation-positives who are healthy at the time of testing go on to develop the disease. As all disease subjects are mutation-positive, the sensitivity is 1.00, but the low penetrance of C282Y means it has a very low predictive value of 0.05 (only 5% of mutation positives develop the disease). These values are not favorable for a population screening program, and the US Preventive Services Task Force recommended against screening of asymptomatic individuals for the presence of the C282Y mutation. Their reasoning was that the identification of a large number of individuals who possess the risk genotype but will never manifest the clinical disease may result in unnecessary surveillance, labeling, invasive work-up, anxiety, and potentially unnecessary treatment. This example shows that it is not only statistical parameters that are taken into account in decisions about genetic risk profiling. There is also an evaluation of the benefits and risks both to the individual and society.

Newborn screening is effective

The discussion so far refers to the identification of persons at high risk of developing complex diseases and should not be confused with the effectiveness of the **newborn screening** routinely done in developed countries. In newborn screening, mutations with very high penetrance that are necessary and sufficient, or deterministic, for monogenic disease development are examined. An example is phenylketonuria (PKU), which is very severe if not prevented, but easily preventable with modification of diet. Newborn screening tests (using either metabolic or genetic markers) are well established and cost effective in diagnosing babies born with such diseases before they develop any symptoms. Today, more than 2000 genetic tests are currently in use for rare deterministic mutations whose presence shows absolute correlation with the development of monogenic diseases. Screening in order to predict risk of a complex disease, for which multiple nondeterministic and low-penetrance variants cause the disease collectively and interact with environmental factors, is a very different proposition.

Effective risk markers have been found for adverse drug effects

While prediction of disease risk has proven difficult, promising results have been obtained in pharmacogenetic studies in which markers for an adverse drug effect show very high sensitivity and specificity. For example, individuals who possess *HLA-B*57* as the risk marker will almost always develop an adverse effect to anti-HIV treatment with abacavir, and those who are negative for *HLA-B*57* will never suffer from the reaction. In this case, classification is complete and the test separates susceptible from nonsusceptible individuals. Several other deterministic markers have been described for adverse effects and, in fact, some have been approved for clinical use by the US Food and Drug Administration (FDA).

Most pharmacogenetic markers of adverse drug effects are HLA antigens, which are involved in the peptide presentation that initiates an immune response. The effectiveness of these markers shows that small molecules, such as drugs or metabolites, can modify the immune response and cause adverse effects. This information has also been exploited to pave the way for new treatments for some autoimmune diseases. A few experimental small molecules are in clinical trials to investigate their effectiveness at preventing or treating autoimmune disorders by interfering with the presentation of autoantigens that triggers the disorder. Pharmacogenetic studies therefore not only provide the strong markers for certain traits, but can also lead the way in offering new treatments by helping to elaborate the mechanisms behind genetic associations.

Genetic risk is measured by a number of parameters

A case-control study generates an odds ratio for a population. Genetic risk profiling concerns the individual, and some measure of genetic risk beyond the OR is preferable. One challenge faced by genetic counselors in their practice is that a patient does not take a risk described in relative terms as an answer, but demands to know whether they will have the disease or not. For complex diseases with a nondeterministic genetic background, such a definitive answer may never be possible, but it may be possible to get close to an answer.

Disease risk is measured by a number of parameters in epidemiology and these are applicable to genetic risk estimation as well. The conventional parameters listed in **Table 11.2** estimate the impact of genetic polymorphisms on disease risk in a population, but they do not provide individual-level information with certainty.

Table 11.2 Measures of risk used in conventional epidemiology

	Definition	Calculation	Usage
Absolute risk	Overall probability of developing a disease	Incidence rate of disease in a population (corresponding to the individual probability of developing the disease)	More meaningful than relative risk (or odds ratio) in health policy decisions
Relative risk	Change in disease risk due to a specific risk factor. Not an estimate of individual-level risk	The ratio of incidence of disease in people with the risk marker to incidence in people without the risk marker	Indicates the strength of the association (effect size) in cohort studies. Independent of the risk marker frequency
Odds ratio (relative odds)	Change in disease risk due to a specific risk factor (an approximation to relative risk). Not an estimate of individual-level risk	The ratio of the odds of disease in people with the risk marker to the odds in people without the risk marker	Indicates the strength of the association (effect size) in case-control studies. Independent of the risk marker frequency
Population attributable fraction	Amount of risk due to the risk marker's presence in a population	The ratio of the difference in incidence rates with and without the risk marker in the population to the incidence rate. Heavily influenced by risk marker frequency	Indicates the impact on the population of removal of the risk marker. Can be high if the marker is common despite a very low odds ratio
Heritability	The proportion of phenotypic variance that is the result of genetic variation	The ratio of genetic risk variance to total risk variance	By indicating the portion of risk variation due to genetic variation, it informs about the genetic background to disease susceptibility

Note that the *P* value is not a risk measure and is not listed. Population attributable fraction (PAF) is included in the table but is not relevant to individual genetic risk prediction. PAF is often misinterpreted as it tends to be high for markers with a high frequency even though their association is weak. A high PAF should not be interpreted as showing that most of the underlying genetic cause of disease risk has been identified. A more important point is how much of the variation in disease risk is explained by the genetic markers, a value which is quantified by heritability. The risk markers for diseases with high heritability will be easier to identify and will have greater predictive ability. Type 1 and type 2 diabetes (high and low heritability, approximately 0.75 and 0.25, respectively) are contrasting examples for the role played by heritability in the ability to predict disease risk. The greater the heritability and the proportion of heritability explained by genetic risk markers (as in type 1 diabetes), the more powerful the risk markers will be in predicting disease development. Success in genetic risk profiling is therefore more likely for diseases with high heritability.

Genetic risk markers for complex diseases are always present in some healthy people and therefore have low predictive ability for future disease development in currently healthy people. The aim of risk profiling is to have a genetic marker that can be used for effective prevention—a marker that can identify persons at risk without having to screen an excessive number of people in order to catch one person at risk. The parameter **number needed to screen (NNS)** quantifies the number of people that need to be tested to detect one person truly at risk of the disease. A good genetic marker should have a low NNS to have utility in complex diseases. Conventional measures of risk are only useful for an initial assessment of markers. They are insufficient for definitive conclusions on the predictive value of markers because genetic risk profiling must provide additional value over existing risk models based on nongenetic risk markers. This is assessed by:

- The additional incremental value that the genetic marker adds to the existing prediction model based on nongenetic risk factors

- Its clinical utility, which is the magnitude of the contribution to the change in predicted risk to warrant a revision in current clinical practice

- Its cost effectiveness

11.2 Assessing Clinical Utility

A future predictive genetic test is first identified as a risk marker in the initial replicated association study, which yields a *P* value and an odds ratio. From the data available, further parameters are calculated to assess the usefulness of the marker in classifying individuals as at risk of developing a disease. If these parameters are favorable, further requirements must be fulfilled before the marker can be declared as a predictor for future development of the disease and treated as a **biomarker**.

A marker must be validated and assessed as clinically useful before it can be considered for clinical use

Any marker identified at the discovery and replication phase as having potential future clinical use is then analytically validated, and it must also be clinically validated and tested for clinical utility (**Figure 11.2**). The additional phases of development are crucial for the test to obtain regulatory approval for clinical use. The discovery phase is the initial

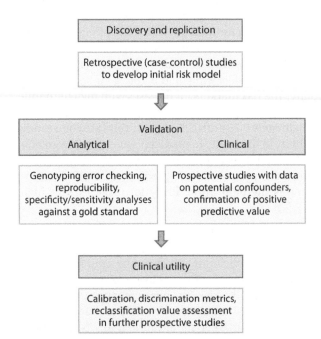

Figure 11.2 The steps involved in confirmation of a marker as a biomarker for genetic risk profiling.

association study, which is usually a small case-control study that includes people who have already developed the disease and those who are disease-free. Analytic validation is straightforward for SNP-type genetic markers, as modern genotyping methods are well established and generally robust (Chapter 7). Analytic validation also measures the sensitivity and specificity of a marker against a gold standard.

For clinical validation, statistically more powerful, and preferentially prospective, studies including a baseline cohort of healthy people are ideal. Avoidance of extensive subgroup analysis and an overfitted (or overadjusted) statistical risk model is equally important. Such efforts may generate a model but it will not be reproducible. The clinical validity step confirms the power of the marker to show correlations with the desired endpoint, which may be development of the disease, response to a drug treatment, or prognosis. In this step, potential confounders (such as age, sex, or race/ethnicity) are also considered. This phase can be seen as yet another replication stage using a more sophisticated approach than a case-control study. Clinical validity studies establish the **positive predictive value** (**PPV**) of the marker, which ideally should be 100%. The positive predictive value answers the question of what proportion of people positive for the predictive marker actually develop the trait. Likewise, the **negative predictive value** (**NPV**) should also be 100% in an ideal test. NPV is assessed in the prospective study by confirming the proportion of people without the predictive marker who do not develop the trait.

Even after confirmation of clinical validity (similar results in the initial case-control study and follow-up prospective study, no evidence of confounding, and high positive and negative predictive values), further clinical utility tests must be performed before the marker can be used in clinics. These tests are performed in further prospective studies and serve to **calibrate** the initial risk model generated in the case-control study and assess the marker's ability to **reclassify** subjects.

Sensitivity and specificity are better measures of a marker's ability to classify individuals than the *P* value and OR

Some of the high ORs reported in case-control studies compare risk at the extreme ends of risk distribution (highest quartile versus lowest quartile). The situation is very different when a currently healthy sample of the population is screened for future individual risk estimation, where such high ORs do not translate into a good prediction. Sensitivity is a measure of the proportion of people with the disease who have the marker, and specificity indicates the proportion of people without the disease who do not have the marker. The marker in the example presented in **Table 11.1** is present in only 5% of cases (50 out of 1000) and therefore has a sensitivity of 5% (0.05). It is absent in 95% of cases and so lack of the marker does not mean lack of disease. However, the marker is absent in all disease-free controls and therefore has a specificity of 100% (1.00). Ideally, both values should be 100%, but this is only possible for a few specific mutations in monogenic traits. Sensitivity and specificity should therefore be as close to 100% as possible in order to be able to distinguish between cases and controls. Several established markers currently in clinical use, including prostate-specific antigen (PSA) for prostate cancer, are judged by their sensitivity and specificity values for their clinical utility.

The **likelihood ratio** is a useful parameter derived from combining sensitivity and specificity and corresponds to the ratio of the true-positive rate to false-positive rate. It is obtained by dividing the sensitivity value by (1 − specificity) and can take any value up to infinity. Thus, the higher the likelihood ratio, the better is the predictive ability of the marker.

Likelihood ratio goes up as both sensitivity and specificity approach their maximum value and its upper limit is infinity. It is therefore most useful for comparing two markers in order to choose the better one, rather than as an absolute measure of clinical usefulness.

PPV and NPV are also useful measures of accuracy

Like sensitivity and specificity, PPV and NPV are measures of diagnostic test accuracy, and both should also be 100% for an ideal predictive marker. In the example presented in **Figure 11.3**, all individuals positive for the marker develop the disease, so the PPV is 100% (1.0); only 51% of individuals negative for the marker do not develop disease, so the NPV is 51% (0.51). PPV and NPV vary with the frequency of a disease in a population (prevalence); a test with a given sensitivity and specificity can have different PPV and NPV in different populations.

ROC, AUC, and C-statistics test clinical utility

All the parameters explained thus far are individually useful for different public health or clinical purposes, but for genetic risk profiling, a parameter is needed that provides cumulative information to quantify the predictive value of a genetic marker within a scale.

Sensitivity and specificity are combined into a **receiver-operating-characteristic (ROC) curve**, which can be used to calculate the **area under the curve** (AUC). An ROC curve plots sensitivity against (1 − specificity) for the marker(s) being tested, and it is the AUC that reflects the probability that the person with the marker(s) being tested will develop the disease (**Figure 11.4**). This probability is also called the concordance- or C-statistics and provides the best information for the classifier property of the marker. It is a discrimination statistic to discriminate those at risk from those not at risk. The C-statistics ranges from 0.5 (the marker has no discriminatory value for people at risk or not) to a maximum of 1.0 (100% discriminatory value). A high value therefore suggests good classification ability of the marker to distinguish between those who will develop the disease and those who will not. The C-statistics for PSA levels in ROC curve analysis is 0.68 for prostate cancer versus no cancer. Given that a test that performs no better than chance yields a C-statistics of 0.50, 0.68 does not suggest a high clinical utility in classifying subjects as

	Disease develops	Disease does not develop	Row totals		
Marker-positive	50	0	50	PPV	50/50 = 1.00
Marker-negative	950	1000	1950	NPV	1000/1950 = 0.51
Column totals	1000	1000			
	Sensitivity	Specificity			
	50/1000 = 0.05	1000/1000 = 1.00			

Figure 11.3 Sensitivity, specificity, positive predictive value (PPV), and negative predictive value (NPV) calculations in a prospective study. Note that the numbers are per 1000 subjects who have developed a disease, and per 1000 subjects who have not developed the same disease at the end of a prospective study, and are not from a cross-sectional sample of a population (thus, the data do not reflect the prevalence of the disease).

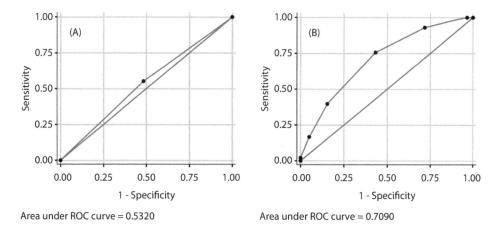

Area under ROC curve = 0.5320   Area under ROC curve = 0.7090

Figure 11.4 Receiver-operating-characteristic (ROC) curves estimate the ability of markers to discriminate between two groups. The diagonal straight line indicates the baseline discrimination that can be obtained by random allocation of people to the two groups (for example, cases and controls), and corresponds to an area under the curve (AUC) of 0.50. (A) A single marker with negligible discrimination ability generates a ROC curve with an AUC of 0.53. (B) A multimarker set increases the AUC from 0.50 to 0.71, and obviously has greater ability to discriminate cases from controls. An ideal marker would have generated an AUC value of 1.0.

having prostate cancer or not. Although there is currently nothing better for the early diagnosis of prostate cancer, the routine use of PSA levels for population screening purposes has been heavily criticized in recent years.

Studies based on empirical data have shown that to achieve a C-statistics value of a modest 0.75, the proportion of genetic variance that a test must explain in the statistical analysis of the association study must be between 0.10 and 0.74. A single GWAS top hit usually explains no more than a few percent of the genetic variation playing a role in disease development. We are therefore still a long way from having satisfactory predictive tests based on current GWAS results. A C-statistics value of 0.75 is the threshold for a diagnostic classifier to be considered clinically useful when applied to a sample at increased risk. The threshold value for a diagnostic classifier to be applied in the general population is 0.99, and no genetic test is currently even close to this threshold.

Calibration of the initial risk model is vital for accurate prediction of risk

Ultimately, accurate prediction of actual risk by the C-statistics depends on the calibration of the risk model originally generated in the case-control study—the level of agreement between predicted and true risk. Calibration of the initial model requires another prospective study. In this confirmatory study, the predicted risk model is tested and compared with the observed true risk. The model may need to be rescaled upward or downward depending on the observed risk, which is called calibration. In the case of a calibration failure, the actual predictive value of the markers varies from the value predicted at the clinical validity phase (say, instead of 50%, 20% of marker-positive individuals develop the expected trait in the follow-up).

Reclassification is crucial in the assessment of clinical utility

Confirmatory prospective studies are large, independent studies that use formal statistical tools to assess a marker's expected clinical utility. The statistical tools used are different from those that assess associations in case-control studies. A marker becomes a biomarker once its clinical utility is confirmed, meaning that it provides additional information that is useful in making clinical decisions about disease susceptibility, progression, or prognosis over and above existing risk markers. Thus, clinical utility is more about what improvement in predictive ability the new marker provides rather than what new information it provides.

The emphasis in assessing clinical utility is the reclassification of subjects, using the new marker, from where they would be classified using existing evidence. The existing evidence usually consists of nongenetic factors and already known genetic factors. Reclassification analysis assesses whether the model including the new markers changes a person's risk sufficiently to move them to a different risk category. A marker that has a modest effect or no effect by C-statistics may improve risk classification at the individual level.

Multiple markers can be used to increase accuracy

Single markers are unlikely to be considered for clinical use unless they yield an unusually high OR (>200). Multiple markers with small effect sizes may be used simultaneously to generate high effect sizes. A simulation study that examined 40 independent genetic risk markers, each yielding an OR of 1.2 to 2.0 in a sample size of one million, concluded that the best results in discriminative accuracy quantified by the C-statistics would be obtained when all risk genotypes were common (≥30%) and ORs for each marker were all closer to 2.0. Even then, the C-statistics was 0.93. This example shows that even the simultaneous use of the most common and strongest risk markers may still not have the desired discriminatory accuracy (≥0.99) between disease and healthy states. Other studies indicate that inclusion of all markers in the model has less power to increase the predictive accuracy than excluding noncausal markers. Therefore, it is important to remove statistical noise and only consider causal markers when using multiple markers to increase the accuracy of prediction.

11.3 Disease Biology and Pharmacogenomics

GWASs have identified more than 2000 risk markers in more than 300 complex diseases at the genome-wide statistical threshold ($P < 5 \times 10^{-8}$) but, on average, odds ratios are less than 1.5. More crucially, identified markers account for a modest fraction of heritability (generally, up to 30%) in complex diseases. These are not promising figures for the development of robust predictive tests, but the development of predictive tests is not the only aim of GWASs (see Figure 11.1).

GWASs have been most useful in the identification of disease development pathways. Even genomic regions with no genes have yielded strong markers of susceptibility and their mode of action has been worked out—for example, a long-range interaction with a proto-oncogene for cancer risk markers. More obvious findings include modification of the action of specific genes, most of which effects were previously unsuspected (**Table 11.3**).

Besides a better understanding of the pathophysiology of specific diseases, GWASs have yielded additional information about genome biology. The revelation that the majority of disease associations are with SNPs located within noncoding regions of the genome shows that the noncoding parts of the genome are not necessarily "junk" DNA and are actually active in regulating gene expression. The efforts directed toward elaborating the function

Table 11.3 Selected examples of novel biological insight into disease biology obtained exclusively from GWAS results

Disease	Gene	Mechanism
Age-related macular degeneration	*CFH*	Complement activation
Crohn disease	*ATG16L1*	Autophagy; bacterial defense
Coronary artery disease	*CDKN2A, CDKN2B*	Cell cycle regulation
Type 2 diabetes	*CDKAL1* *MTNR1B*	Cell cycle regulation Circadian rhythm
Childhood nonallergic asthma	*ORMDL3, GSMDB*	Dysregulation of the unfolded protein response in airways; the effect of dysregulated sphingolipid synthesis on bronchial hyperreactivity
Obesity	*FTO*	Disruption of inhibitory control of food intake via long-range interaction with *IRX3* and *RPGRIP1L*

of noncoding genomic regions have revealed how genetic variation correlates with non-coding RNA function, epigenetic changes (both at the DNA and chromatin level), and with functions of genes on different chromosomes. Overall, the notion that it is primarily coding-region variants that affect disease susceptibility has been transformed into the recognition that noncoding regions are at least as equally instrumental as coding regions in the mediation of disease susceptibility.

Pharmacogenetics

While complex disease genetics are difficult to unravel due to the involvement of many variants with small effects, it appears that both drug metabolism and adverse reactions are regulated by relatively few variants that can have a big impact on clinical management. Genetic association studies, mainly GWASs, have yielded important clues for drug development by identifying genes important in drug metabolism pathways. It does not matter that the odds ratios for such effects are small as they still provide sufficient information on which genes are involved in the pathways and are potential targets. The first example was the observation that HMG-CoA reductase was involved in cholesterol metabolism, which eventually led to the development of statins as the major lipid-lowering drugs. The next generation of lipid-lowering drugs will target the gene *PCSK9* and these drug-development efforts were initiated by observations that sequence variants in this gene are associated with lower lipid levels and protection from cardiovascular diseases. Currently, efforts are underway to introduce new diabetes treatments that target the peroxisome proliferator-activated receptor-γ (*PPARG*) based on the identification of this gene's variants as modifiers of type 2 diabetes, albeit with very small effect sizes.

Future directions

From early candidate gene studies to the GWAS era, genetic association studies have contributed plenty of information to the better understanding of disease biology and to drug development efforts, as well as identifying some markers that may be used in differential diagnosis. The GWAS era began with high expectations and some may think that it has not lived up to these expectations. However, given the insights provided to genome biology and disease pathophysiology, and the detection of genetic markers already in clinical use following their approval by the US FDA for the identification of individuals susceptible to potentially lethal adverse drug reactions, it would be unfair to conclude that GWASs have not fulfilled their promise. We also have to remember that the first GWAS was published in 2006 and it typically takes more than a decade for a translational finding to reach the clinics. Thus, while there is still room for improvement in the development of predictive tests for complex diseases, the advances in the understanding of disease biology and pharmacogenetics that have been brought to us by GWASs are already making a difference in the way medicine is practiced.

The GWAS may not have uncovered all genetic susceptibility markers for complex diseases, but it has provided enough information to bring genomics to clinics. It has accomplished its role in initiating a genomic medicine era, which will proceed even faster with the arrival of affordable next-generation sequencing. The developments in the post-GWAS era—namely, the new sequencing technologies and advances in epigenomics—will further accelerate the translation of gene discoveries to improved health for individuals and populations.

Key Points

- Neither an extremely small *P* value nor a very high odds ratio is an indicator that a marker is a good classifier of individuals for their disease risk category. Thus, identification of a risk factor does not necessarily provide a strong basis for development of a biomarker.

- The area under the ROC curve or C-statistics incorporates sensitivity and specificity of genetic testing and is a useful metric for discrimination of risk.

- Even when population-based risk parameters are impressive, especially with the use of multi-locus risk markers, at the individual level, genetic markers have not led to much progress over existing nongenetic factors as predictors of risk.

- A marker becomes a biomarker once its clinical utility is confirmed, meaning that it provides information that is useful in making clinical decisions about disease susceptibility over and above that provided by existing risk markers.

- The usefulness of genomic risk prediction depends on the genetic architecture (such as the total number and frequencies of risk markers, and heritability) and prevalence of the disease, as well as the proportion of the variation in disease attributable to genetics. Thus, success is more likely for diseases with high heritability, low prevalence, and the presence of common risk markers with high odds ratios.

- An ideal biomarker for disease susceptibility has high sensitivity and specificity (positive in every predisposed subject, and negative in every subject that will remain disease-free), is well calibrated (the predicted risk for developing the disease is not over- or underestimated), and will classify every subject as predisposed or not. This scenario is currently only true for some monogenic Mendelian disorders where the genetic effect is not modified by the environment.

- GWAS results have not provided strong predictors of complex disease risk, but have delivered novel insights into disease biology that are useful for drug development and personalized medicine.

- The GWAS has been a giant step in genomic medicine but should be seen as only the first step to achieving the translational aims of genomic medicine. The developments in the post-GWAS era, particularly the new sequencing technologies together with epigenomics, should allow the wider use of genomic medicine.

Further Reading

Abraham G & Inouye M (2015) Genomic risk prediction of complex human disease and its clinical application. *Curr Opin Genet Dev* 33, 10–16 (doi: 10.1016/j.gde.2015.06.005).

Attia J, Ioannidis JP, Thakkinstian A et al. (2009) How to use an article about genetic association: C: What are the results and will they help me in caring for my patients? *JAMA* 301, 304–308 (doi: 10.1001/jama.2008.993).

Becker F, van El CG, Ibarreta D et al. (2011) Genetic testing and common disorders in a public health framework: how to assess relevance and possibilities. Background Document to the ESHG recommendations on genetic testing and common disorders. *Eur J Hum Genet* 19(Suppl 1), S6–S44 (doi: 10.1038/ejhg.2010.249). (*An extensive discussion of complex disease genetics, susceptibility markers, quantitative measures of risk, genetic testing and population screening, and steps of biomarker development including economic assessment and cost-effectiveness analysis. The current status of genetic screening for specific complex diseases is also reviewed. It also discusses ethical, legal, and sociological issues associated with genetic testing and screening.*)

Bodmer W & Bonilla C (2008) Common and rare variants in multifactorial susceptibility to common diseases. *Nat Genet* 40, 695–701 (doi: 10.1038/ng.f.136).

Chatterjee N, Wheeler B, Sampson J et al. (2013) Projecting the performance of risk prediction based on polygenic analyses of genome-wide association studies. *Nat Genet* 45, 400–405 (doi: 10.1038/ng.2579). (*The utility of polygenic models for risk prediction will depend on achievable sample sizes for the training data set, the underlying genetic architecture, and the inclusion of information on other risk factors, including family history.*)

Jakobsdottir J, Gorin MB, Conley YP et al. (2009) Interpretation of genetic association studies: markers with replicated highly significant odds ratios may be poor classifiers. *PLoS Genet* 5, e1000337 (doi: 10.1371/journal.pgen.1000337).

Janssens AC, Ioannidis JP, van Duijn CM et al. (GRIPS Group) (2011) Strengthening the reporting of Genetic Risk Prediction Studies: the GRIPS statement. *PLoS Med* 8, e1000420 (doi: 10.1371/journal.pmed.1000420).

Janssens AC, Moonesinghe R, Yang Q et al. (2007) The impact of genotype frequencies on the clinical validity of genomic profiling for predicting common chronic diseases. *Genet Med* 9, 528–535 (doi: 10.1097/GIM.0b013e31812eece0).

Jostins L & Barrett JC (2011) Genetic risk prediction in complex disease. *Hum Mol Genet* 20(R2), R182–R188 (doi: 10.1093/hmg/ddr378).

Kooperberg C, LeBlanc M & Obenchain V (2010) Risk prediction using genome-wide association studies. *Genet Epidemiol* 34, 643–652 (doi: 10.1002/gepi.20509).

Kraft P, Wacholder S, Cornelis MC et al. (2009) Beyond odds ratios—communicating disease risk based on genetic profiles. *Nat Rev Genet* 10, 264–269 (doi:10.1038/nrg2516).

Manolio TA (2013) Bringing genome-wide association findings into clinical use. *Nat Rev Genet* 14, 549–558 (doi: 10.1038/nrg3523).

Glossary

1000 Genomes (1KG) Project: An international project aimed at obtaining the whole genome sequence of 1000 subjects to find most genetic variants that have frequencies of at least 1% in the populations studied. Currently, open access data from 2577 subjects are available.

2 × 2 table: A table with two rows and two columns, which is typically used to summarize case-control study results as counts (rows: case and control counts; columns: marker-positive and -negative counts).

3′ untranslated region (UTR): The transcribed but not translated part of a gene at the end of the last exon. May contain key regulatory elements that are involved in post-transcriptional regulation of expression of the message through influencing the stability of mRNA and also through microRNA binding.

5′ untranslated region (UTR): The transcribed but not translated part of a gene at the beginning of the first exon. May contain regulatory sequences involved in regulation of transcription.

Additive model: The genetic model that examines a linear increase in the risk with increasing numbers of the risk allele in the genotype. If the risk for heterozygotes is increased "r" times, and "r + r" times for rare allele homozygosity, the risk change fits in the additive model. This analysis yields an odds ratio per allele.

Admixture mapping: Mapping a disease gene by taking advantage of a recently admixed population where ancestral populations have differences in frequencies of disease-associated alleles. It exploits the fact that the segments in the chromosomes from the ancestral population with higher disease frequency will contain an increased density of ancestry informative markers.

Allele: Alternative nucleotides at a position that make up the single nucleotide polymorphism; or, more broadly, alternative versions of a gene.

Allelic heterogeneity: The situation where multiple mutations in the same gene may result in the same phenotype (disease).

Allelic model: See multiplicative genetic risk model.

Alternative explanation: Explanations other than the one offered. Usually means consideration of chance, bias, and confounding in the interpretation of epidemiologic study results.

Alternative splicing: Formation of diverse mRNAs through differential splicing of the same RNA precursor, which may result in multiple peptides from a single protein-coding gene (or differences in 3′ UTR length).

Ancestry informative marker (AIM): A genetic marker that shows large differences in frequency across population groups. These loci are useful in ancestry determination. These markers can be used in admixture mapping or admixture matching in case-control studies.

Area under the curve (AUC): The area under a receiver-operating-characteristic curve (range 0 to 100%), which plots sensitivity against (1 − specificity) of a test, mainly for comparison of different cutoff values or with other biomarkers (or tests) for their discrimination value.

Association: Tendency of a marker and a trait to occur together more often than expected by chance. It is a statistical correlation between the marker and trait, and does not mean causation.

Bias: Deviation of the observed results from true results, or processes leading to such deviations. Most bias can be prevented by good research design. Bias is one of the main alternative explanations for results in an epidemiologic study.

Bias toward the null: Any bias that results in a deviation in the results toward no association when the true result is an association.

Biochemically manifesting heterozygotes: Heterozygotes for a recessive disease-causing mutation (carriers) with biochemical changes in the absence of clinical findings.

Biological interaction: An interaction between two (risk) factors with a known biological mechanism, which results in the risk conferred by one differing in the presence of the other one.

Biological plausibility: The agreement of the mechanistic inferences from the observed association with accepted biologic processes.

Biological significance: The significance of a result in biological terms; whether or not a finding results in a significant change in a biological process.

Biomarker: A biological marker such as a genetic variant that can be measured accurately and reproducibly, and can predict the occurrence of a disease or its outcome.

Binary variable: A variable with two possible values (0 or 1; case or control; yes or no; present or absent).

Bioinformatics: A discipline within computational molecular biology that deals with understanding and analyzing complex biological phenomena.

BioMart facility: A search engine designed to explore data within large databases using comprehensive queries.

Bonferroni correction: A conservative method proposed as a safeguard against false positives arising from multiple comparisons. Not favored when the number of comparisons is too large, as the false-negativity rate increases.

Bottleneck: A sharp decrease in population size, usually due to environmental catastrophes, followed by recovery and return to the previous size.

Calibration: A later step in biomarker development. The calibration step assesses how close the predicted risks are to the actual observed risks.

Canalization: The presence of robust pathways that lead to standard phenotypes despite the deletion of some genes.